science is beautiful

DISEASE AND MEDICINE

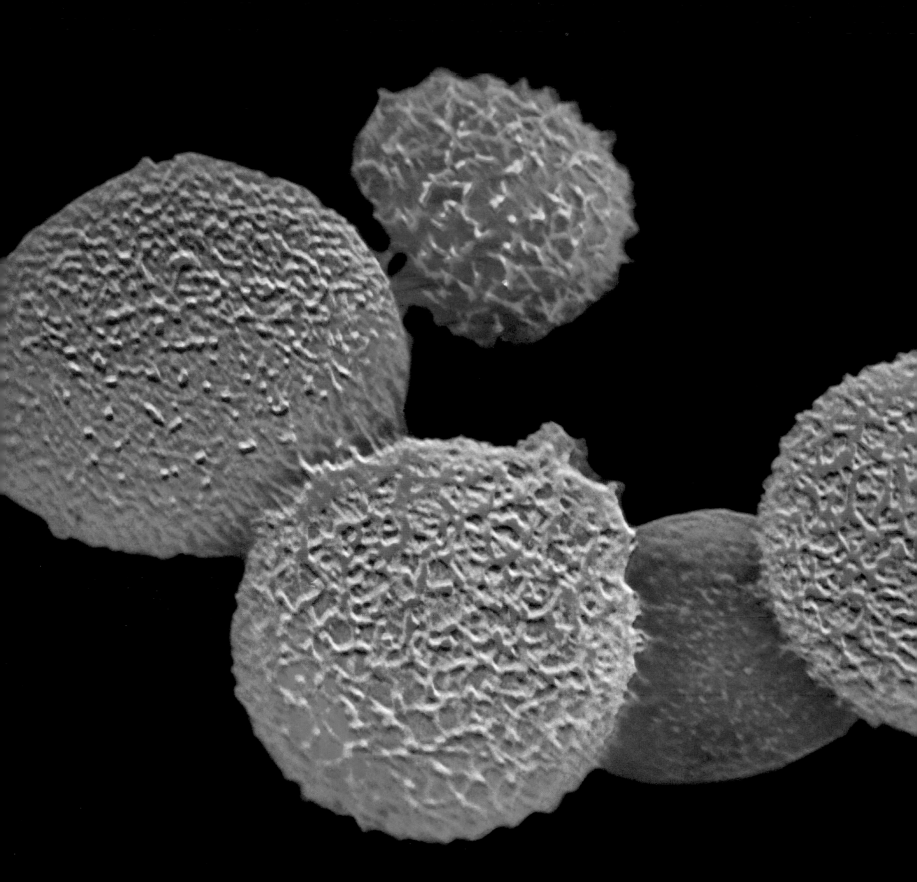

science is beautiful
disease and medicine

UNDER THE MICROSCOPE

Colin Salter

BATSFORD

First published in the United Kingdom in 2017 by
Batsford, an imprint of Pavilion Books Company Limited
43 Great Ormond Street
London WC1N 3HZ
www.pavilionbooks.com

Volume copyright © Batsford

ISBN: 9781849944410

A CIP catalogue record for this book is available from the British Library.

10 9 8 7 6 5 4 3 2 1

Reproduction by Colourdepth, UK
Printed by 1010 Printing International Ltd, China

This book can be ordered direct from the publisher at the website:
www.pavilionbooks.com, or try your local bookshop.

Previous page: Cryptococcus fungi (scanning electron micrograph)
This false-colour image shows several cells of *Cryptococcus neoformans*,
a yeast-like fungus. The cells are covered in a protective casing (here in
green), which preserves them until conditions are favourable for their
activation. The fungi are deposited in the soil via pigeon droppings, and
humans inhale them when the spores become airborne. The disease that
they cause, cryptococcosis, is potentially fatal to those with damaged
immune systems. It is often an indication of the development
of AIDS in patients.
(Magnification x4800 at 6cm wide)

Right: Cocaine crystals (scanning electron micrograph)
Indigenous South Americans had been chewing the leaves of the
coca plant medicinally and recreationally for a thousand years before
cocaine was isolated in a German laboratory in 1855. Now cocaine
is the second most widely used illicit drug in the world after cannabis.
It has a medical application as a local or topical anaesthetic for eye, nose
and mouth surgery, although its function is often replaced nowadays by
less toxic drugs.
(Magnification unknown)

Contents

In the first book in this series, *Science is Beautiful: The Human Body Under The Microscope*, we looked in close-up detail at how the body works. We live in bodies which are masterpieces of complex machinery, a network of interacting, finely tuned systems which in normal use work so well that we don't even have to think about their operation. Our lungs breathe, our hearts beat, our faces crease with sadness or laughter, without any conscious instruction from us, the machines' owner-occupiers.

In this book we're turning our attention to what happens when things go wrong. Specifically we are looking at how bacteria, viruses and other disruptive forces manage to outwit the sophisticated defence mechanisms of our immune system. We look too at how science fights back, counter-attacking the invading diseases with ingenious medicines to prevent or cure disease. Remarkably, disease itself can sometimes be harnessed in the service of health. Doctors may deploy a so-called Trojan virus, which invades cells but prevents the spread of a more virulent infection. Vesicular stomatitis virus, for example, is used in this way to treat sufferers of HIV, cancer and ebola. Vaccination is another approach which actually gives us a small dose of a disease; this triggers the body's defence against a larger dose. The first vaccine, devised in 1796, has been so successful that the disease it protected us from, smallpox, has been completely eradicated from the world.

Finding a Cure

Medicine is big business. It takes a long time for a new drug to be approved for public use, a process which requires significant investment long before any prospect of a financial return. Suppose there's a disease that you'd like to find a cure for: first you have to understand the science of the disease. What does it affect? Where is it acquired? How is it transmitted? Only then can you begin your search for treatment. Once you've found something

you think might work, you may start laboratory tests on cultured, infected cells. Only if the data from those tests looks promising can you start to do clinical trials on actual patients.

If after all that the drug seems to be effective, with side effects which are acceptable when compared to the medical benefits, you can start to produce and advertise your new miracle. Even then, everything has to be tested from the means of production and safety of packaging to the clarity of the leaflet telling you how to take the drug. And at every stage you are regulated by your country's medical authority – for example the Food and Drug Administration (FDA) in the US and the National Institute for Health and Care Excellence (NICE) in the UK.

It all costs a fortune, which is why pharmaceutical companies seek to protect their investment with patents giving them exclusive rights to manufacture and market new drugs. What seems like a high price for a medicine has to cover the costs not just of the pills and the bottle they come in but of years of research and development. But such high prices naturally discriminate against poor people and poor countries who may need the drugs most of all.

Fighting the Cure

Successful medicines bring their own challenges. The increasing resistance of bacteria to the antibiotics used to treat them has been a problem since the 1950s. Bacteria evolve to develop their resistance; but sometimes the medicine itself is to blame. Antibiotics may kill off the weaker strains of a bacterium, allowing the stronger, more resistant ones to thrive. And indiscriminate antibiotic medicines may destroy not only the bad bacteria but the good – bacteria that helps our digestive system for example. The general public still thinks of antibiotics as almost miraculous cure-alls, and reaches out for them to deal with any illness. But no amount of antibiotics will successfully treat a virus-borne disease; and our misuse of antibiotics for such infections is another factor in their decreasing effectiveness.

Sometimes our own immune system turns on us. Autoimmune illnesses such as rheumatoid arthritis, multiple sclerosis, Crohn's disease and ulcerative colitis begin when the body's defences mistakenly act against healthy cells. Transplanted organs may also trigger this response, if the immune system decides to reject them. These unwelcome immune responses are countered with medicines that suppress them – immunosuppressant drugs. The trouble is, they suppress the immune system, leaving us vulnerable to diseases which we would normally be able to repel.

Acquired immune deficiency syndrome, AIDS, has the same effect. It's not a disease itself but a condition that develops from prolonged infection by human immunodeficiency virus, HIV. Patients with AIDS are prone to infections against which a healthy immune system would easily defend. The increasing incidence of AIDS, and the growing use of immunosuppressant drugs, are being blamed in some quarters for the spread of some diseases such as MRSA that were previously unable to get much of a hold on the human population.

Cat and Mouse

There is a constant game of cat and mouse between science and nature. Viruses are simpler, less intelligent organisms than us; but just like us they are programmed to survive, and will adapt to do so. Vaccination against influenza is a running battle because of the disease's ability to mutate into strains against which existing vaccines are ineffective. You may draw some comfort from the fact that it is not usually in the virus's interest to kill its host. to do so would eliminate its means of reproduction and continuation as a species.

Instead, many viruses simply give us a bad cough or dose of diarrhoea which sends them on through the air or the sewage system to infect someone else. If meanwhile science finds a way to defeat them through vaccination or medication, many viruses will evolve, requiring science to find new vaccines, forcing the virus to evolve... and so on.

Some diseases are more serious. Some remain fatal until, we hope, science finds a cure. But thanks to constant progress in medical care, even without cures, it is now possible for the patients of many once-fatal diseases to live with their illness instead of dying of it. AIDS is the obvious example.

Prevention, as the saying goes, is better than cure. Preventative medicine can be anything from vaccination to vitamin supplements. Maintaining the body's defences involves good diet, regular exercise and restorative sleep as much as anything else. Just washing your hands is the single most important preventative act. Coughs and sneezes are not the only things that spread diseases. Infections are transmitted and acquired via bodily fluids of all kinds – vapour in the breath, blood, sweat and other waste products. Poor hygiene is the great enabler of disease, whether its personal care or the safe communal disposal of human waste. Safe sex, with barrier contraception, or no sex at all, plays its part in disease prevention. So too does drinking a tall glass of clean water; the body is more than 50 per cent water by weight, and it's vital to its efficient working to keep it properly hydrated.

This may not be the book to present to your hypochondriac friends – friends who are liable, having read about symptoms one day, to convince themselves that they have them the next. Of one thing we can be reasonably sure: you don't catch diseases by looking at extraordinary photographs of them. If nothing else, this book allows you to study disease and medicine in fascinating detail and from a safe distance. Scientists have captured these images so that you don't have to. On that subject, a brief note about how these pictures were made may be helpful. You'll see alongside each one a note of what sort of micrograph it is. A micrograph is simply a graphic image of microscopic detail, and there are several different ways of producing one. Most of the pictures in this book were acquired by one of two technological marvels.

Light micrographs

A light micrograph is produced by a light microscope. This is the traditional microscope, the one invented in the sixteenth century which uses lenses to magnify a specimen visible under natural or artificial light. When light strikes an object, it is reflected by the surfaces of that object according to the colour, texture and angle of those surfaces. That reflected light reaches the eye, either directly or (in this case) through the lenses of a light microscope. The light is gathered on light-sensitive cells inside the eyeball. The brain processes the information gathered by these cells, information about shape and size as well as colour and texture, in an activity better known to us as Sight. The light microscope sees more or less what the human eye sees, but simply magnifies it.

The microscope became a tool for scientific study in the late seventeenth century, and remains the simplest, low-tech, low-cost way to look at small things. It has changed little in essence in the four hundred years since its invention. The greatest innovations have been in the kind of light used to view specimens. For example, shining polarized light through a sample can reveal particular patterns of colour and structure in the same way that polarized sunglasses can. You can see this to great effect in the images of medicines in this book.

Electron micrographs

At the start of the twentieth century scientists began to develop a high-tech alternative to the light microscope. The first electron microscopes appeared in the 1930s. Instead of beams of light, they use a stream of electrons fired from an electron gun. Instead of lenses they use electromagnets, which can bend beams of electrons in the same way that glass lenses bend light. If the electron beam is dense enough, it becomes possible for the first time to see things in greater detail than with mere light – in other words, to see things that are not visible to the naked eye.

There are two kinds: the transmission electron microscope (TEM) and the scanning electron microscope (SEM). As its name suggests, the electrons from a TEM are transmitted – that is, they pass right through the material being studied. Because they pass through it, they are affected by it, just as light passing through stained glass is affected. It is the way in which the electrons are affected that builds up an image of the material, just as sunlight streaming through a stained-glass window lets us see the full colourful work of the window's designer. The TEM's image is collected on the far side of the material, either by camera or by a fluorescent screen.

By contrast the electrons from an SEM do not pass through the specimen. The SEM fires electrons which scan the specimen in a grid pattern. They interact with the atoms in the material, which then emit other electrons in response. These secondary electrons may be emitted in many directions depending on the shape and composition of the surface. They are detected; and by combining information from these secondary electrons with details of the original electron scan, a scanning electron micrograph is built up.

Because their electrons have to pass through the material, TEMs can only work with very thin samples of material. SEMs can deal with much bulkier material and the resulting images can convey depth of field. TEMs are however capable of greater resolution and magnification. The numbers are unimaginable, but TEMs can reveal details less than 50 picometres (50 trillionths of a metre) in width and magnify them over 50 million times. SEMs can 'see' details one nanometre (1000 picometres) in size and magnify them by up to half a million times. By comparison, an ordinary light microscope only shows detail larger than around 200 nanometres (4000 times larger than the TEM) and provides useful, undistorted magnification only up to two thousand times.

Most of the micrographs you will see in this book are enhanced with additional colour, sometimes called false colour. This makes it easier to see what is being illustrated, and prettier too. Our bodies are not the multi-coloured works of art you see here. But they are works of extraordinary complexity, marvels of biological engineering with impressive defence capabilities. If from time to time those defences are breached by some new infection, medical science works with the body to defeat it. And science, as we know, is beautiful.

Now please wash your hands.

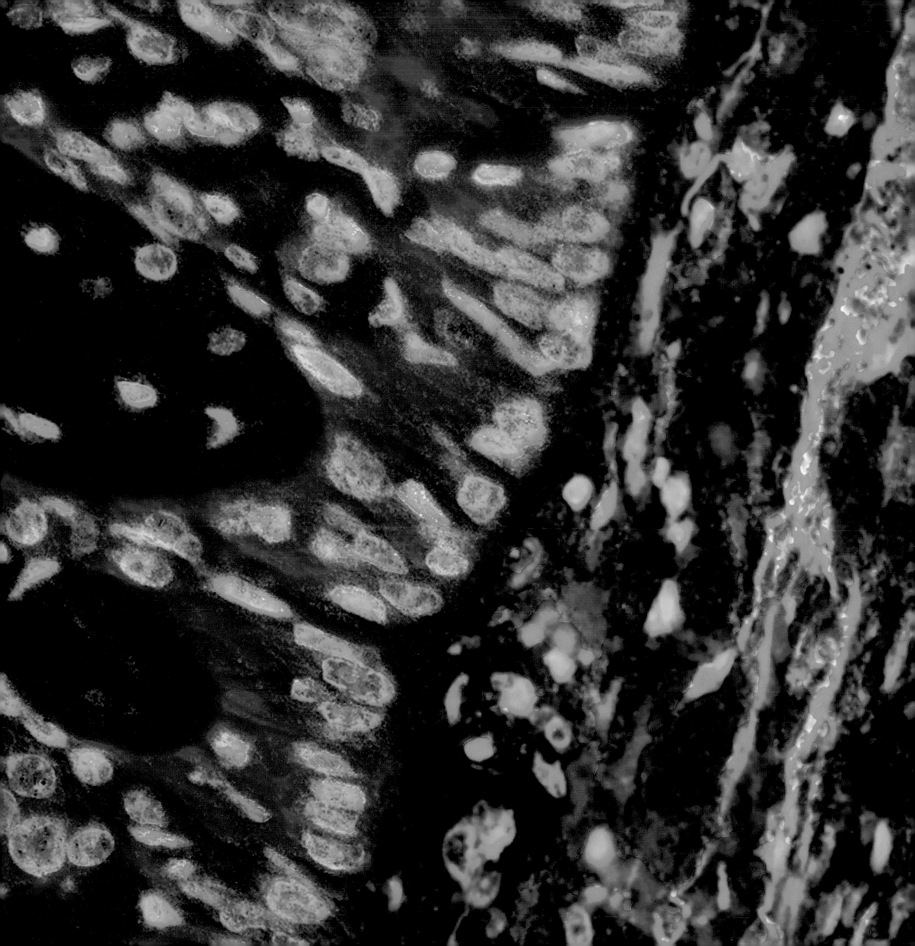

Chronic
Disease

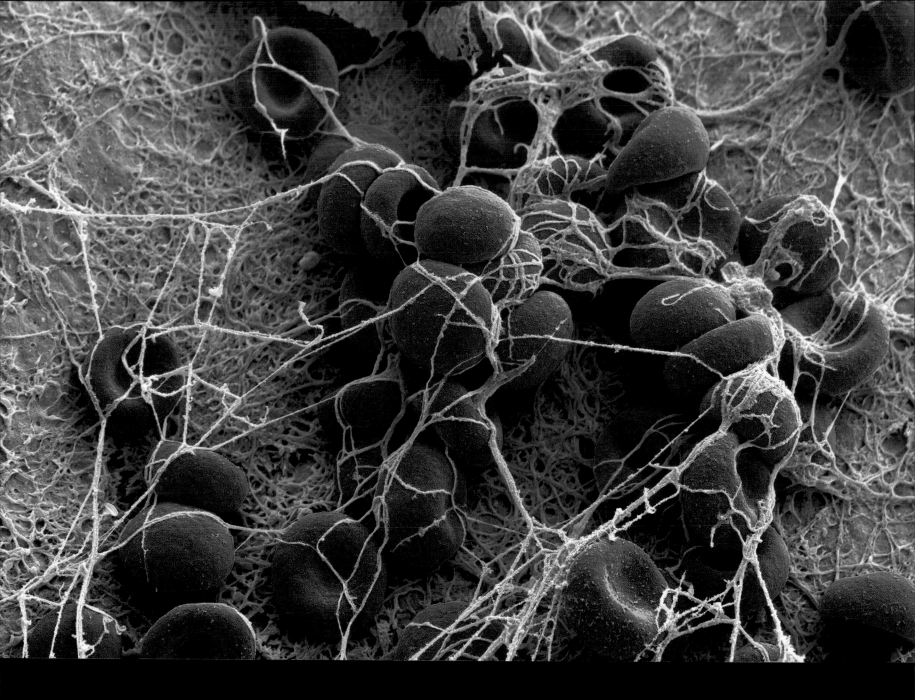

Previous page: Lung cancer desmoplasia (fluorescence light micrograph)
Cancerous lung cells (here red, with blue nuclei) can cause desmoplasia, a fibrous growth in the stroma (connective tissue, here in green) of the lungs, akin to scar tissue. The purple marks are caused by the stroma-specific dye used in this fluorescent light micrograph. The result of desmoplasia is pulmonary fibrosis, a chronic lung condition which makes breathing difficult and therefore severely restricts the activities of sufferers. (Magnification unknown)

Above: Clot forming in lung tissue (scanning electron microscope)
A blood clot or thrombus is formed by the release of a glycoprotein called fibrinogen, the strands of which form a net to capture and retain the escaping blood cells (in red). A clot can form in lung tissue (as in this image), or travel there from elsewhere, often a leg. If it blocks a lung artery it may cause a pulmonary embolism, resulting in chest pain, shortness of breath and coughing up blood. (Magnification unknown)

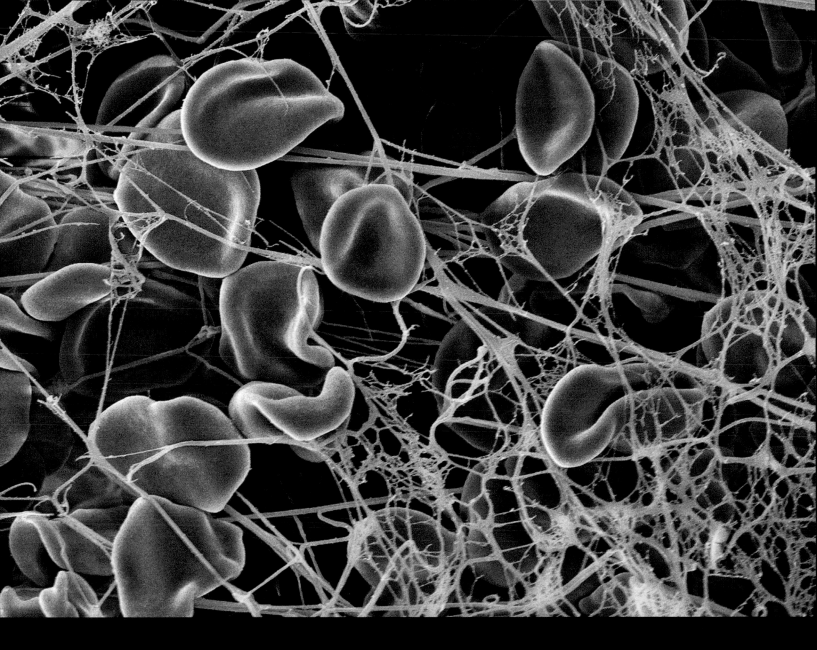

Blood clot (scanning electron microscope)
Both these images, taken by a scanning electron microscope, show tendrils of fibrinogen (also known as fibrin) trapping blood cells to form a clot. The release of fibrin is triggered by platelets in the blood, and normally helps to limit blood loss. Internally however, clots can become detached and form blockages in blood vessels, a common cause of strokes and heart attacks. The risk is reduced by the use of blood-thinning drugs such as heparin and warfarin.
(Magnification unknown)

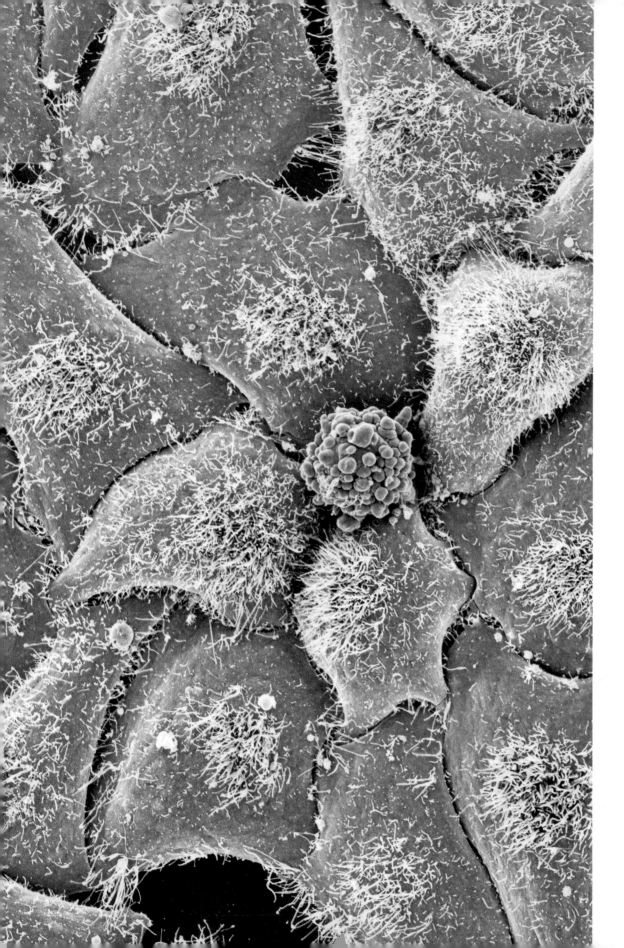

**Left: HeLa cells
(scanning electron micrograph)**

When Henrietta Lack died of cervical cancer, doctors retained a sample of her tumour. They found that its cells (code-named HeLa cells) continued to divide and regenerate in laboratory conditions. This provided researchers with an infinite and consistent source of material on which to experiment with treatments for cancer and many other conditions. In this image the spherical cell at centre is dying, and apoptotic bodies have covered it to prevent toxic substances from contaminating its neighbours.
(Magnification unknown)

Right: HeLa cells (multi-photon fluorescence light micrograph)

Henrietta Lack died in 1951 and by 1954 her immortal cancer cells had been used to develop a polio vaccine. Controversially, the sample of her tumour was removed without her permission. But HeLa cells have since contributed to our understanding of AIDS, radiation and many other conditions, including human sensitivity to everyday materials such as glue and cosmetics. In this image the microtubules of the nuclei are pink, with the core of DNA in blue.
(Magnification x500 at 10cm wide)

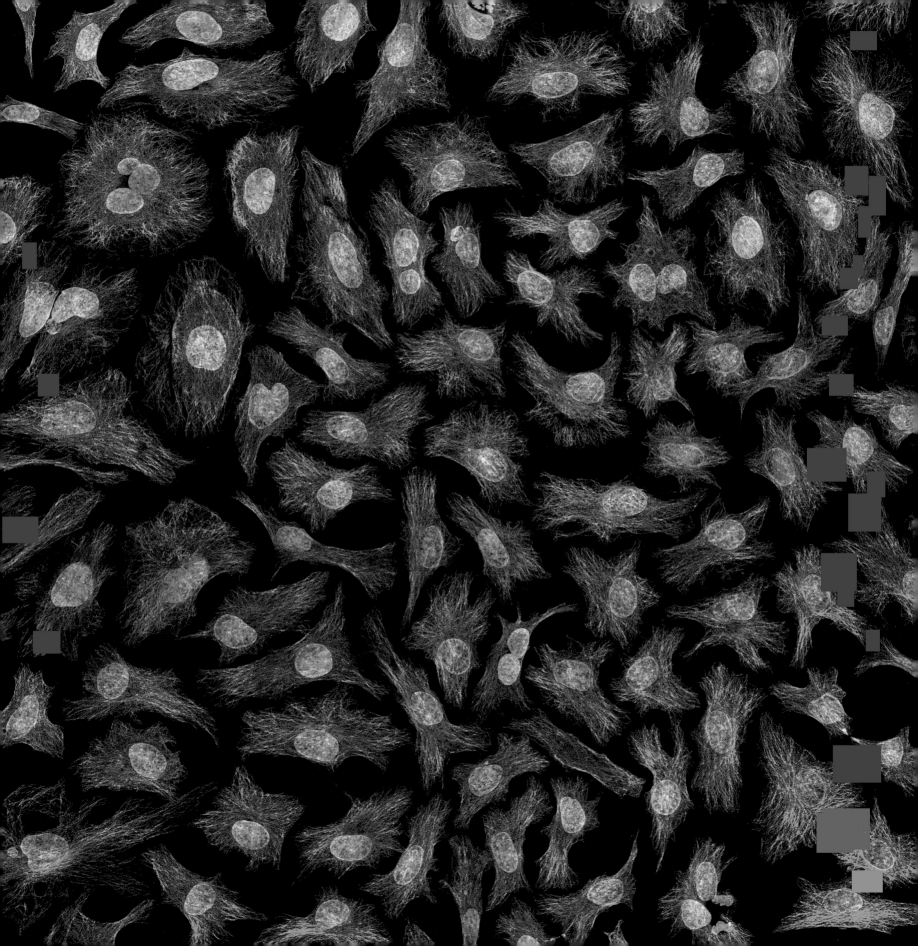

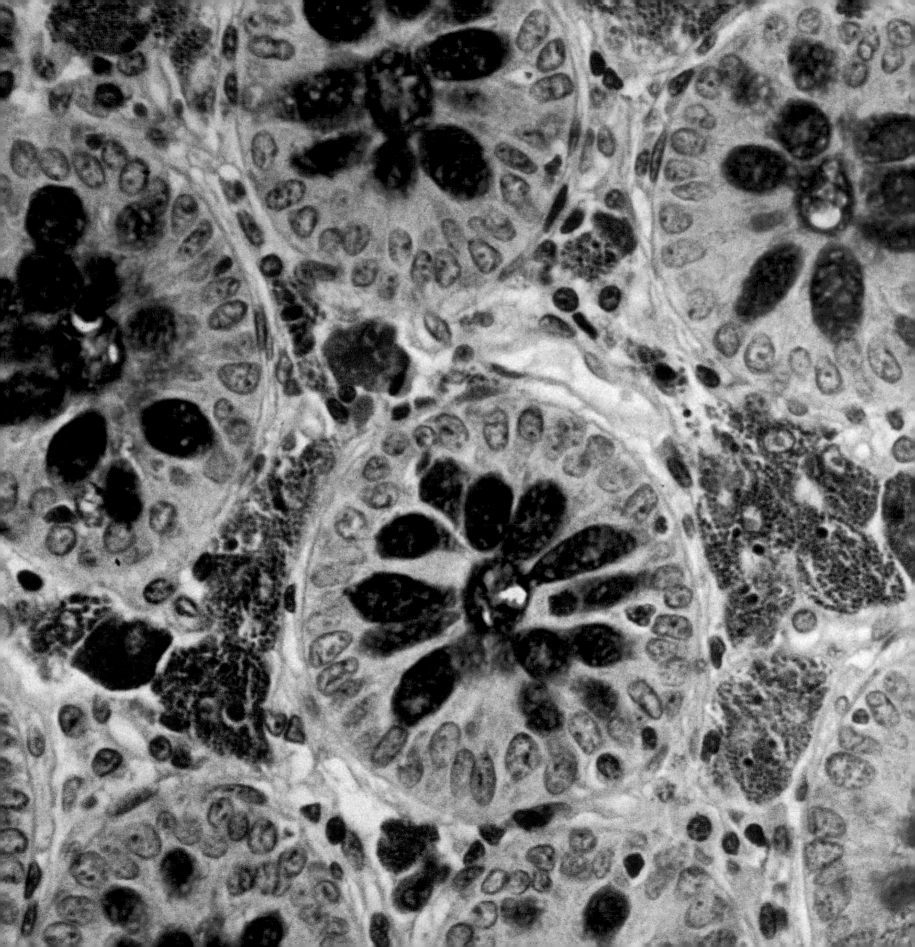

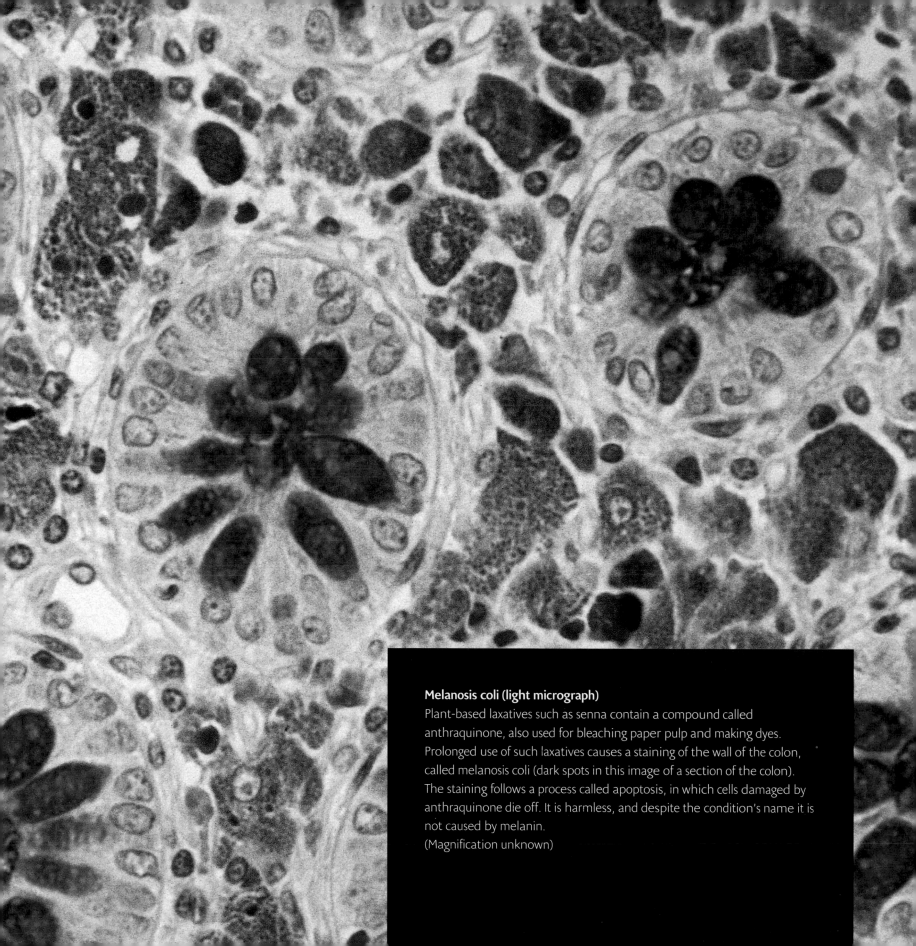

Melanosis coli (light micrograph)
Plant-based laxatives such as senna contain a compound called anthraquinone, also used for bleaching paper pulp and making dyes. Prolonged use of such laxatives causes a staining of the wall of the colon, called melanosis coli (dark spots in this image of a section of the colon). The staining follows a process called apoptosis, in which cells damaged by anthraquinone die off. It is harmless, and despite the condition's name it is not caused by melanin.
(Magnification unknown)

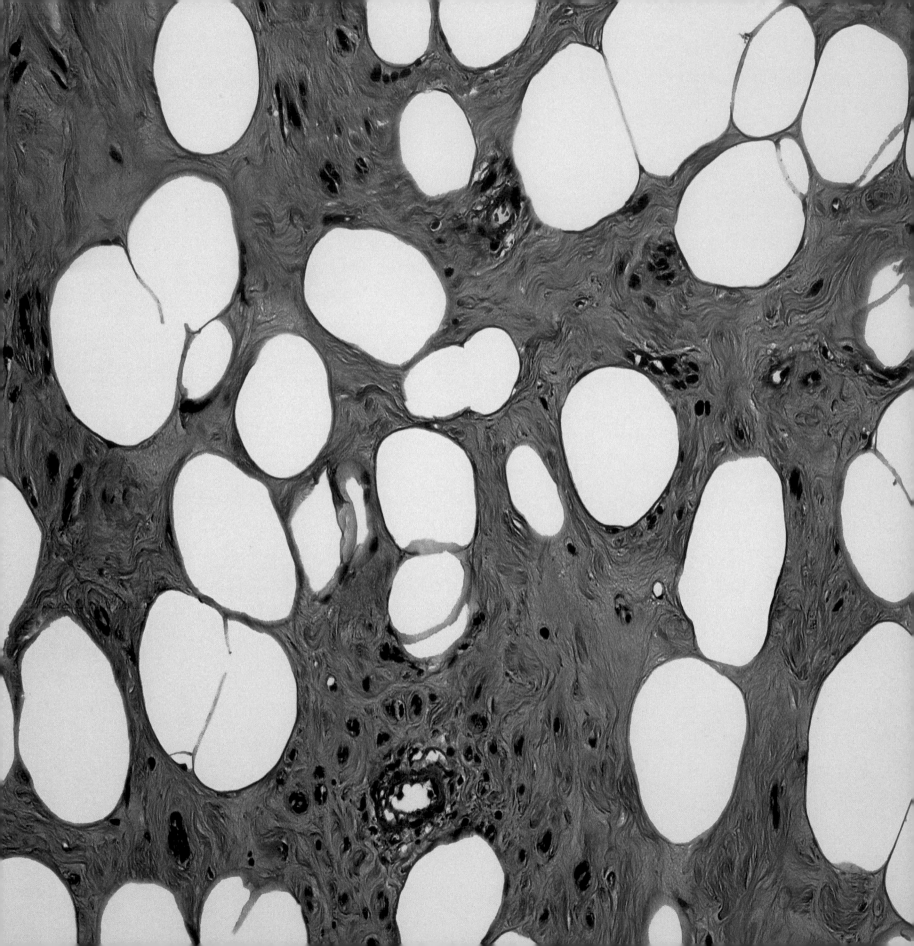

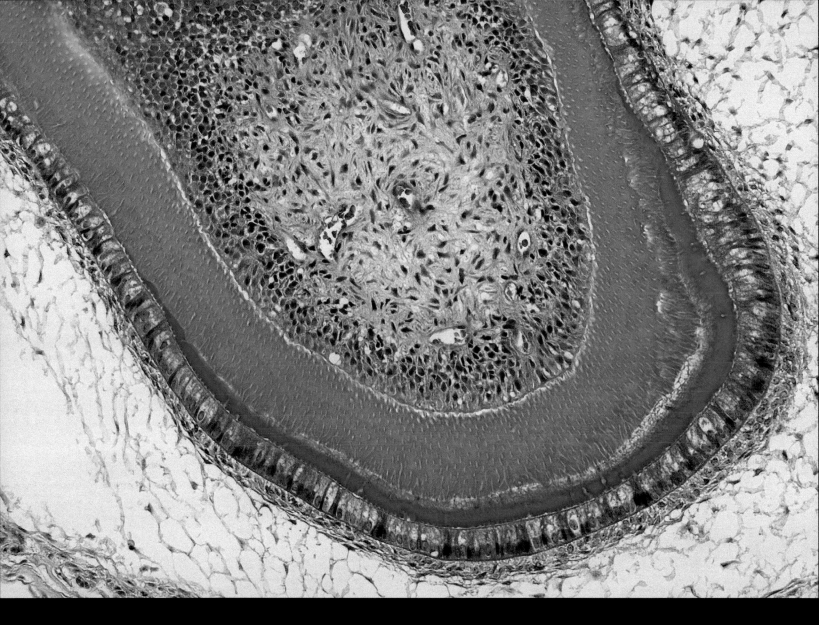

Left: Uterine fibroid (light micrograph)

Uterine fibroids are tumours in the womb consisting of fat cells (adipocytes, pale green here) and smooth muscle tissue (mid blue). Although they are benign, they can cause degrees of discomfort in the lower back or during periods or sex, and may press on the bladder. They often occur in generations of the same family and may be connected to hormone levels. There is no causal connection between them and cancerous uterine tumours.

(Magnification x60 at 10cm wide)

Above: Dermoid ovarian cyst (light micrograph)

This image shows a section through a tooth which has developed within a dermoid ('skinlike') ovarian cyst. Ovaries contain the building blocks for all the cells of the body, so when cysts occur, they may contain bone, hair follicles, sweat glands or even teeth. Although technically benign, they can grow quite long and become twisted, causing abdominal pain. They are generally removed by a simple surgical procedure.

(Magnification x150 at 10cm wide)

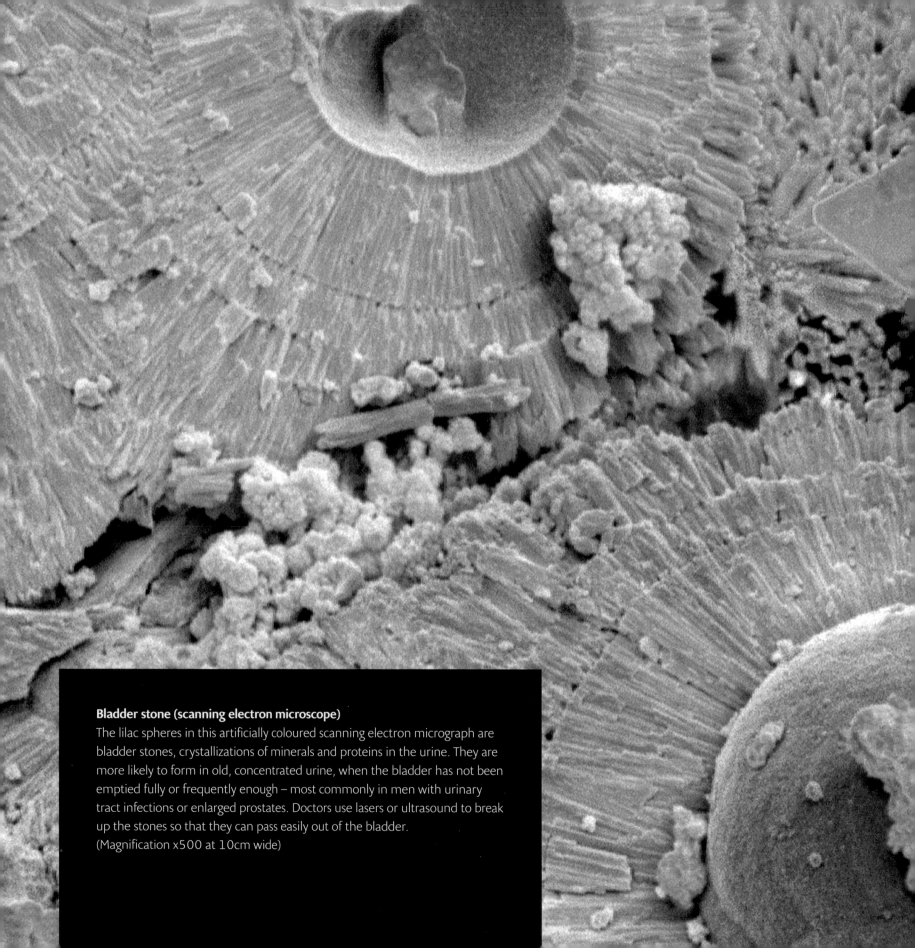

Bladder stone (scanning electron microscope)
The lilac spheres in this artificially coloured scanning electron micrograph are bladder stones, crystallizations of minerals and proteins in the urine. They are more likely to form in old, concentrated urine, when the bladder has not been emptied fully or frequently enough – most commonly in men with urinary tract infections or enlarged prostates. Doctors use lasers or ultrasound to break up the stones so that they can pass easily out of the bladder.
(Magnification x500 at 10cm wide)

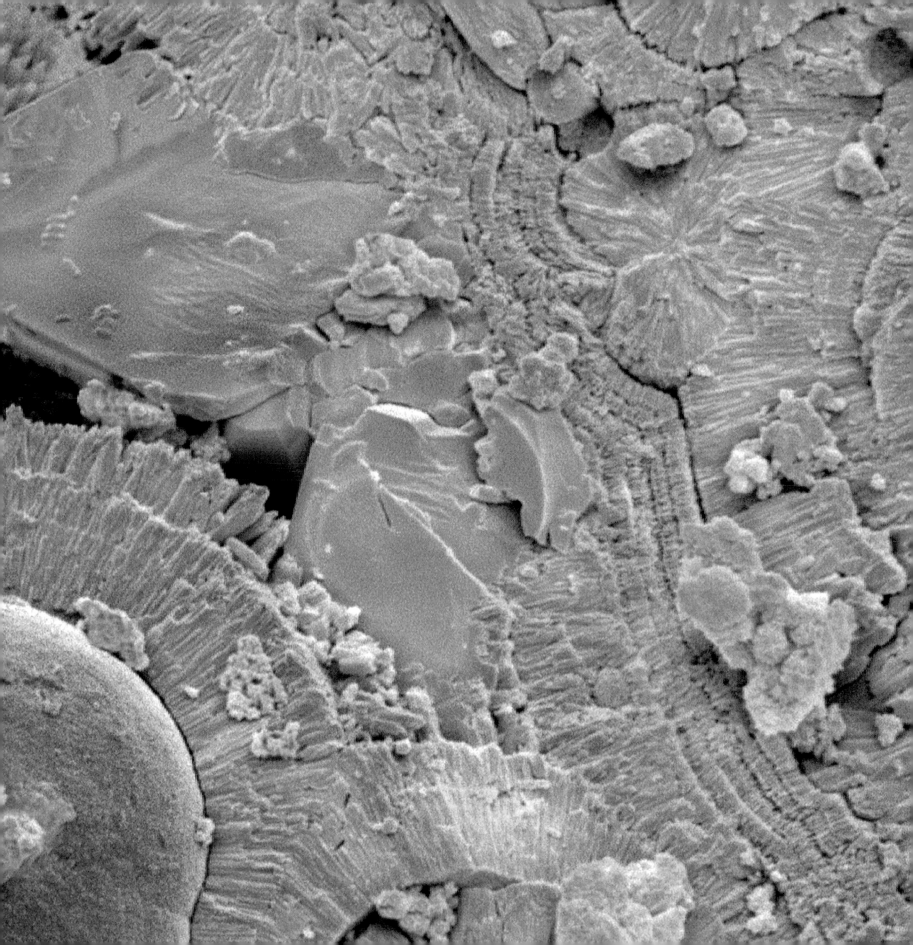

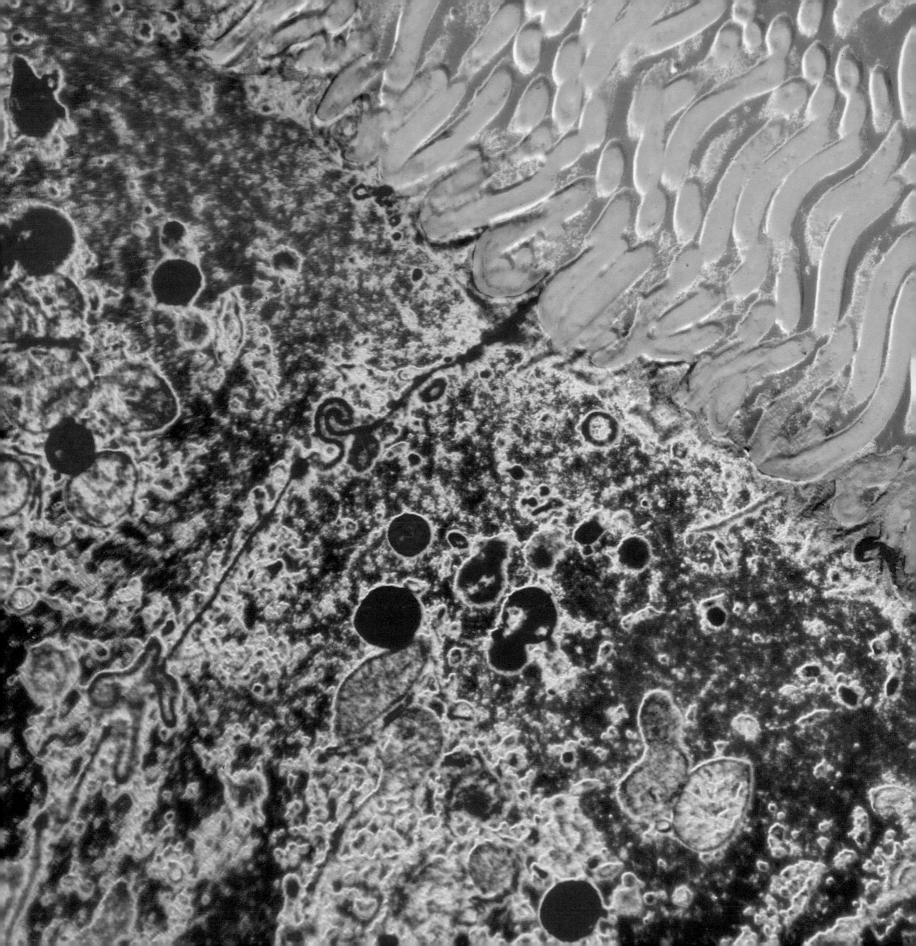

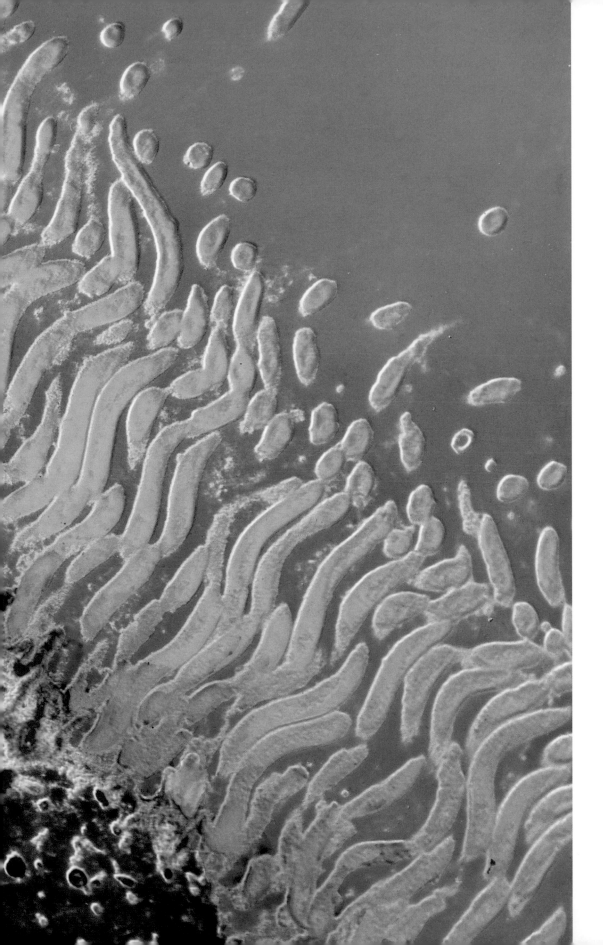

Treponema bacteria in duodenum (oloured transmission electron micrograph)
The pink and orange mass in this image is the lining of the duodenum, the first section of the small intestine. The wormlike yellow strands attached to it are _Treponema_ bacteria. They are spiral in form, and subspecies of _Treponema_ are responsible for one of several disfiguring skin conditions, including pinta, yaws and syphilis. The means of transmission varies for each, and all are treatable with a course of antibiotics, notably pencillin.
(Magnification x4500 at 6x4.5cm size)

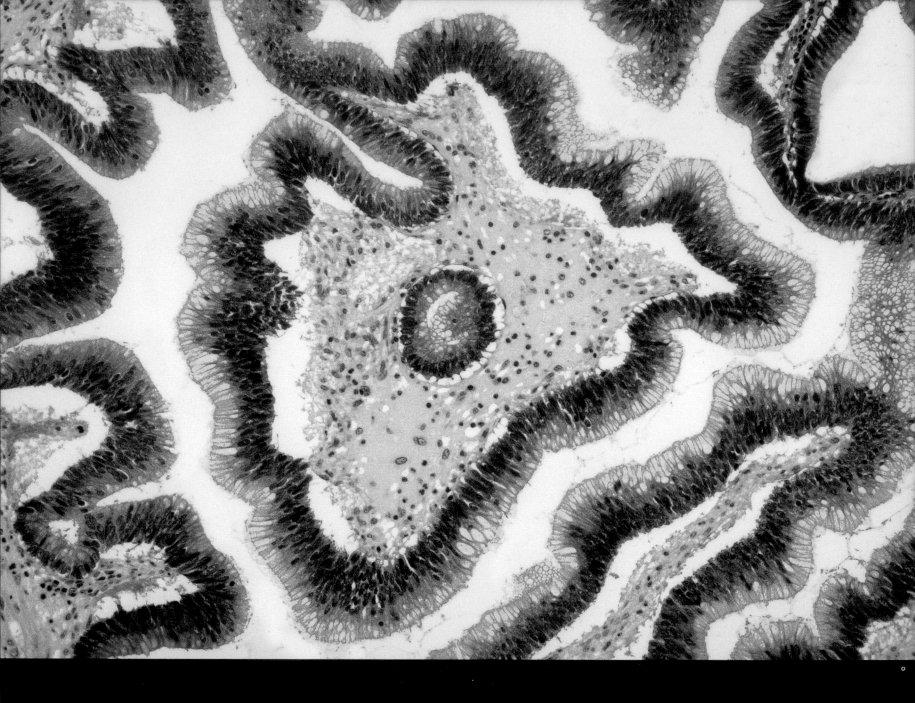

Colon polyp (light micrograph)
This is a cross-section of a villous adenomatous polyp in the colon: that is, a leafy cluster of cells growing from the lining of the bowel. Such polyps may have no symptoms and go unnoticed; but although they are benign at first, they often develop into malignant, cancerous polyps, which can be fatal. They occur most frequently in the over-50s, and regular screening will detect them early enough to be surgically and safely removed. (Magnification x150 at 10cm wide)

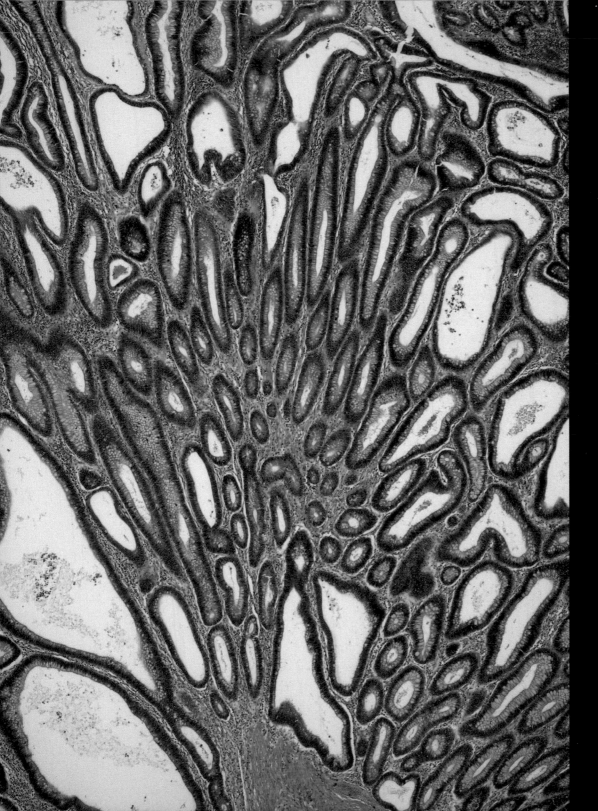

Adenoma of colon (light micrograph)
Adenomatous polyps begin in glandular cells on the inner surface of the colon. They can be either tubular or villous (leafy, as in this image) in form – sometimes a mixture of the two. It is changes in the DNA there that trigger the growth, and can lead to cancer. Poor diet and little exercise are often the cause of these changes, particularly in people over 50. Doctors recommend regular colonoscopies to catch such polyps before they become malignant. (Magnification x80 at 10cm wide)

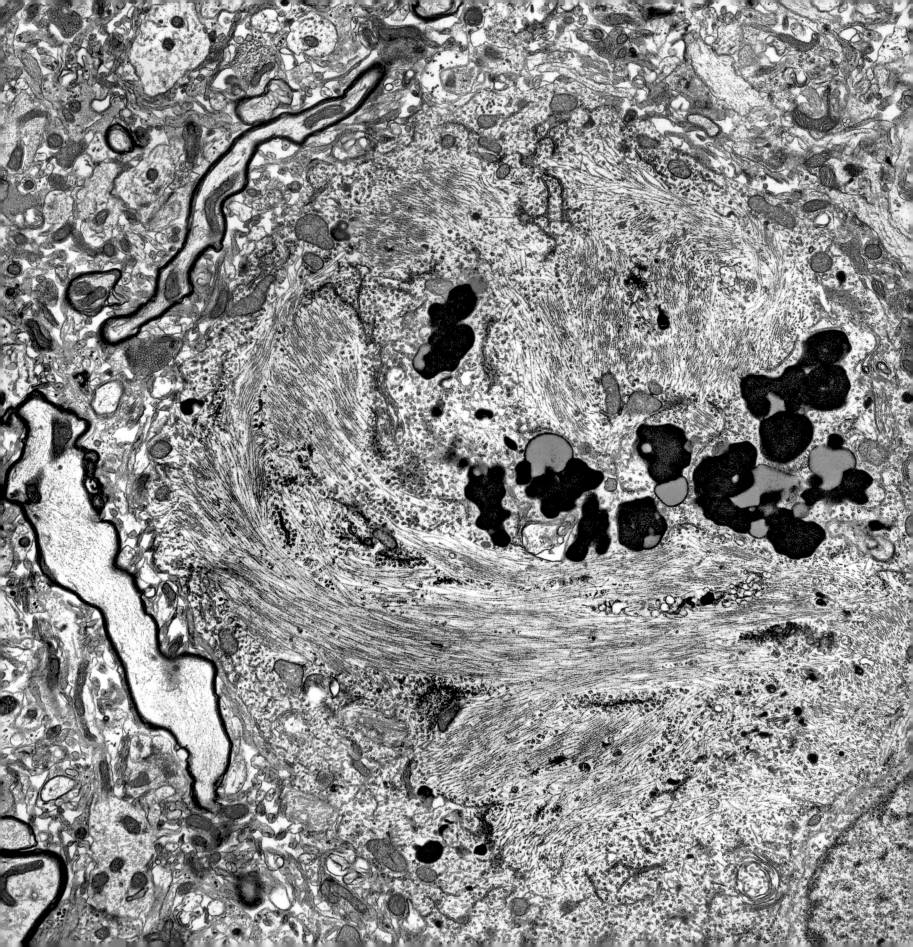

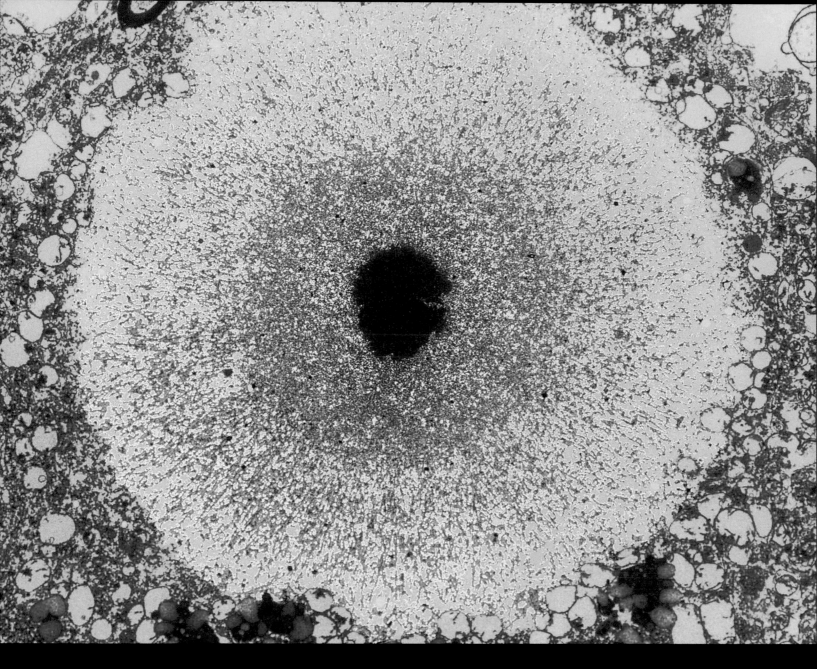

Left: Alzheimer's-disease brain cell (transmission electron micrograph)
In this view of a brain cell affected by Alzheimer's disease, the cell's cytoplasm (the fluid which fills the cell and supports its skeleton) is coloured blue. The swirling brushstroke of green is a tangle of protein fibres, a feature also symptomatic of Creutzfeldt-Jakob disease and other neural disorders. The tangle is caused by an excess of a protein called tau, which in a healthy cell acts to stabilize the cell's skeleton of microtubules. (Magnification x2000 at 10cm wide)

Above: A Lewy body in Parkinson's disease (transmission electron micrograph)
A key diagnostic feature of Parkinson's disease is the presence of so-called Lewy bodies in neurons (nerve cells) within the substantia nigra, the part of the brain which controls movement. Lewy bodies (shown here in blue) are clusters of filaments of a protein called alpha-synuclein, which normally functions to relay communication between neurons. The formation of Lewy bodies inhibits this communication and causes the tremors and restricitons of movement that typify Parkinson's. (Magnification x2750 at 6x7cm size)

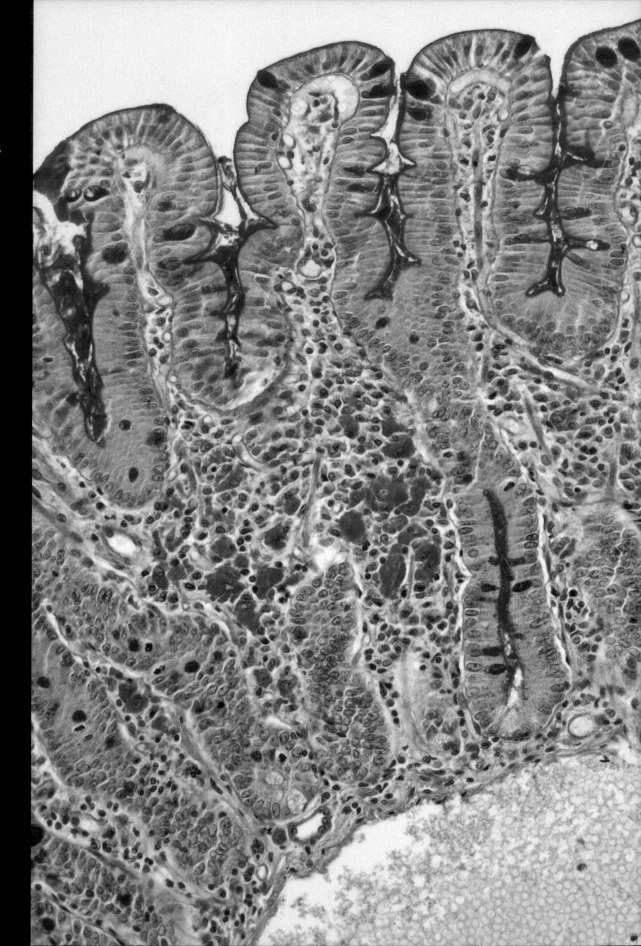

Whipple's disease (light micrograph)

Whipple's disease can affect several organs of the body including the heart and lungs, but most often it strikes the small intestine. It reduces the intestine's ability to absorb nutrition, and symptoms include diarrhoea, weight loss and fatigue. It's caused by the bacterium *Tropheryma whipplei*, seen here as small dark foamy bubbles within the lining of the intestine. It is fatal if untreated, but can usually be overcome with a long-term course of antibiotics.
(Magnification x560 at 10cm wide)

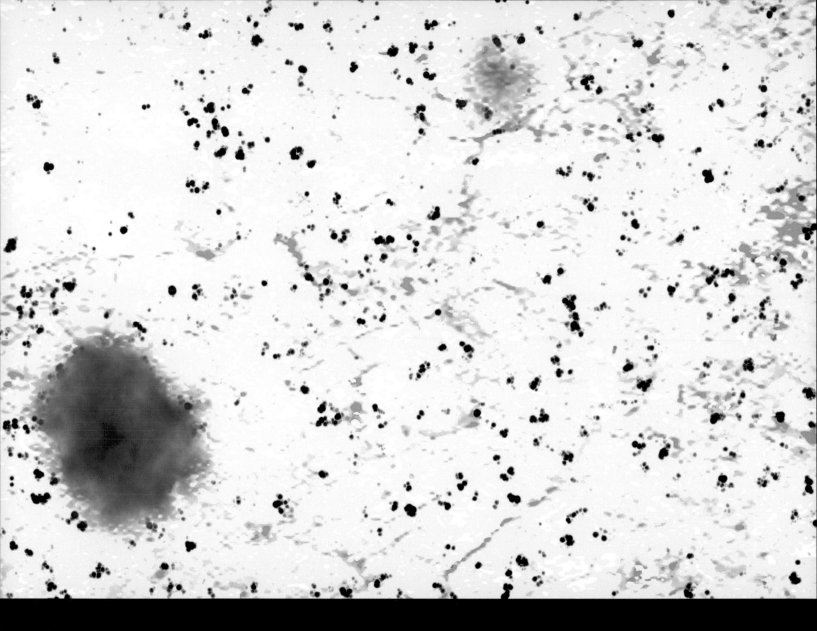

Creutzfeldt-Jakob diseased brain (light micrograph)

Creutzfeldt-Jakob disease (CJD) can occur spontaneously in the elderly, but more often it is acquired through transplants from an infected individual, or by eating beef infected by the animal disease bovine spongiform encephalopathy (BSE). This image shows the sponge-like appearance of a CJD-infected brain: many empty white spaces within the yellow brain matter, surrounding the nerve cells (in red). With neural communication disrupted in this way, CJD sufferers experience memory loss, dementia and involuntary movements.

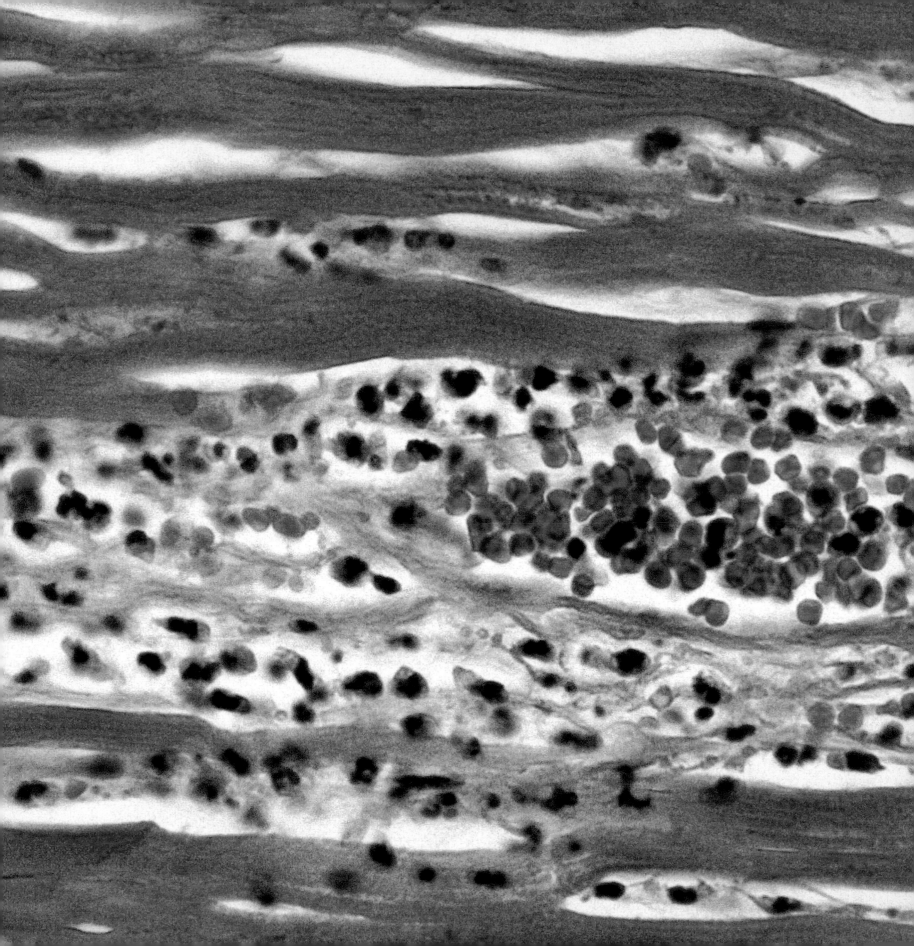

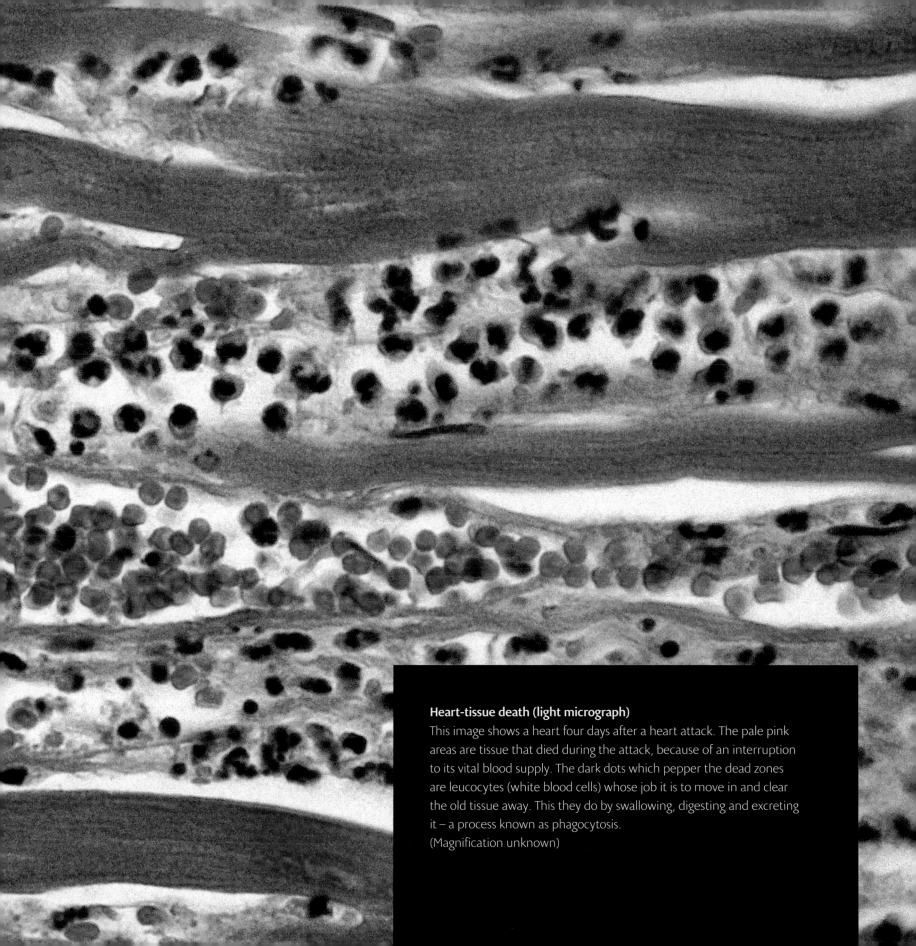

Heart-tissue death (light micrograph)
This image shows a heart four days after a heart attack. The pale pink areas are tissue that died during the attack, because of an interruption to its vital blood supply. The dark dots which pepper the dead zones are leucocytes (white blood cells) whose job it is to move in and clear the old tissue away. This they do by swallowing, digesting and excreting it – a process known as phagocytosis.
(Magnification unknown)

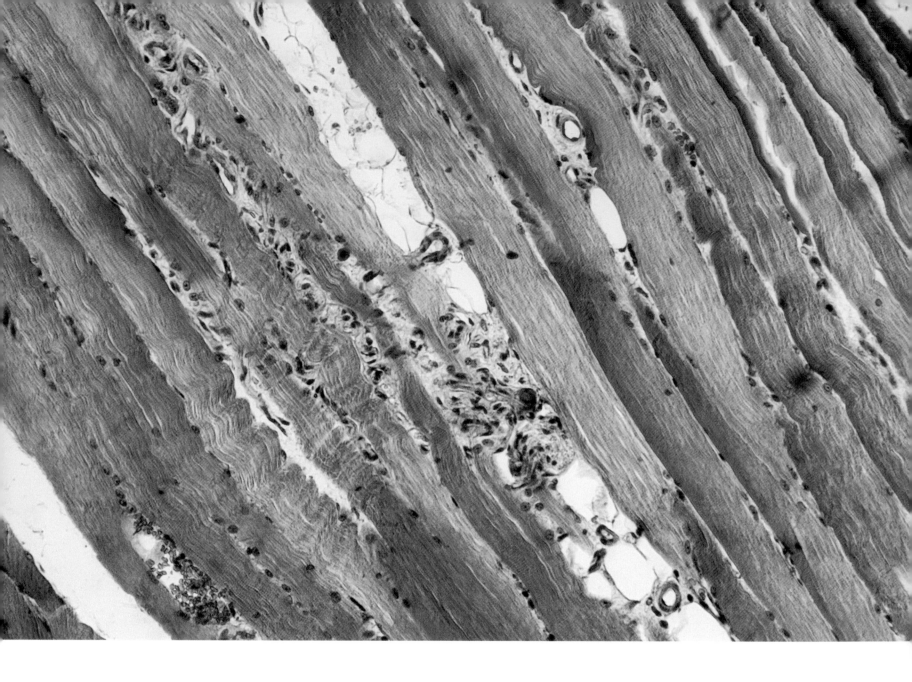

Cushing's syndrome (light micrograph)
Our adrenal glands produce a number of hormones besides adrenaline.
Overproduction of one of them, cortisol, is responsible for Cushing's
syndrome. In this condition, muscles (the pink layers of tissue seen here)
waste away and the patient typically develops a round face and bulges of
fat around the abdomen and between the shoulders (but with thin limbs).
High blood pressure and poor skin condition are also symptoms and the
syndrome is far more common among women.
(Magnification unknown)

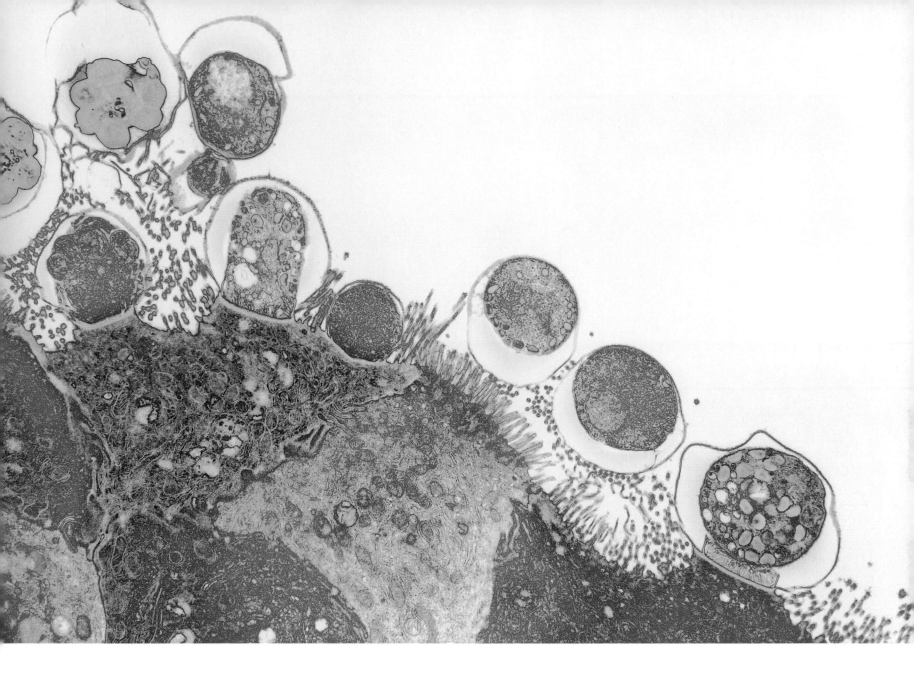

Cryptosporidiosis (transmission electron micrograph)

Cryptosporidium parvum are single-cell parasites (here with their nuclei shown in blue), which are passed to humans from infected milk or water. In this image they are attacking the lining of the intestine (red) where their toxins will cause painful abdominal cramps and severe diarrhoea. Healthy humans can usually overcome the infection with rehydration and regular anti-diarrhoea medicine. But anyone with a weakened immune system will find it harder to resist. Cryptosporidiosis is highly contagious.
(Magnification x2200 at 6x7cm size)

**Kidney stone crystals
(scanning electron micrograph)**
These delicate petals are the crystals of a
kidney stone, formed from calcium oxalate.
The mineral occurs naturally in urine. It can
precipitate and become solid when it is present
in relatively greater concentrations than usual.
This is often the result of diet, dehydration
or hyperparathyroidism. Small kidney stones
pass harmlessly out of the kidneys through
the urinary tract. But larger ones can be
excruciatingly painful and must be broken up
with ultrasound or laser.
(Magnification unknown)

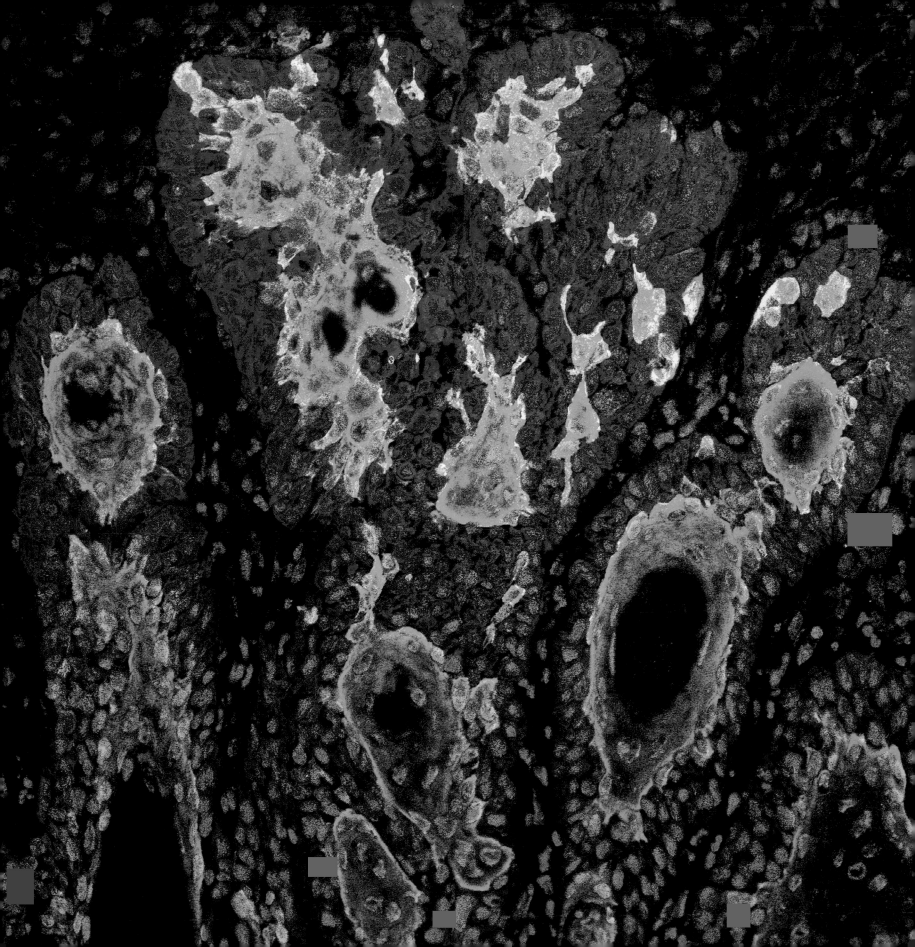

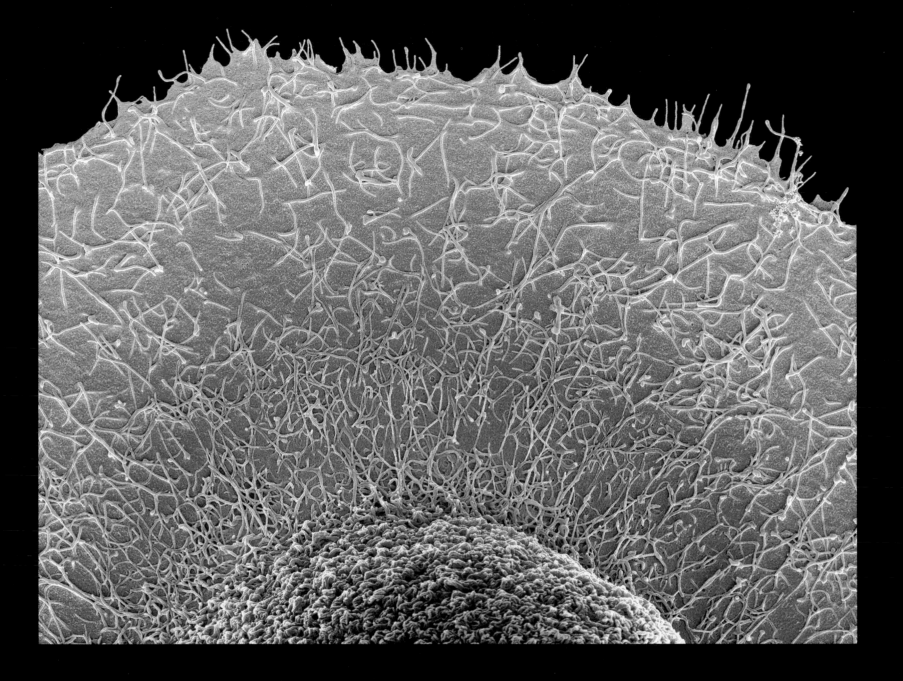

Left: Squamous-cell carcinoma cells (fluorescence light micrograph)
Squamous cells are flat skin cells through which materials are diffused or filtered and where therefore a large surface area is important. They occur in the lungs, mouth, vagina, heart, blood vessels and elsewhere. In this image, the nuclei of squamous cells are blue, and those surrounded by a green wall of keratin are cancerous, connecting with other keratinized cells to form tumours and invading neighbouring tissue. It's one of the commonest forms of skin cancer.
(Magnification unknown)

Above: Colorectal cancer cell (scanning electron micrograph)
This striking artificially coloured image shows part of a cancerous cell in the colon, or large intestine. Colorectal cancer is one of the commonest cancers in the developed world, where smoking, obesity and a diet of red meat and alcohol are often contributory factors. Victims may experience abdominal pain and bleeding from the rectum. Treatment is usually by a combination of surgery and targeted radiation or chemotherapy.
(Magnification x1500 at 10cm wide)

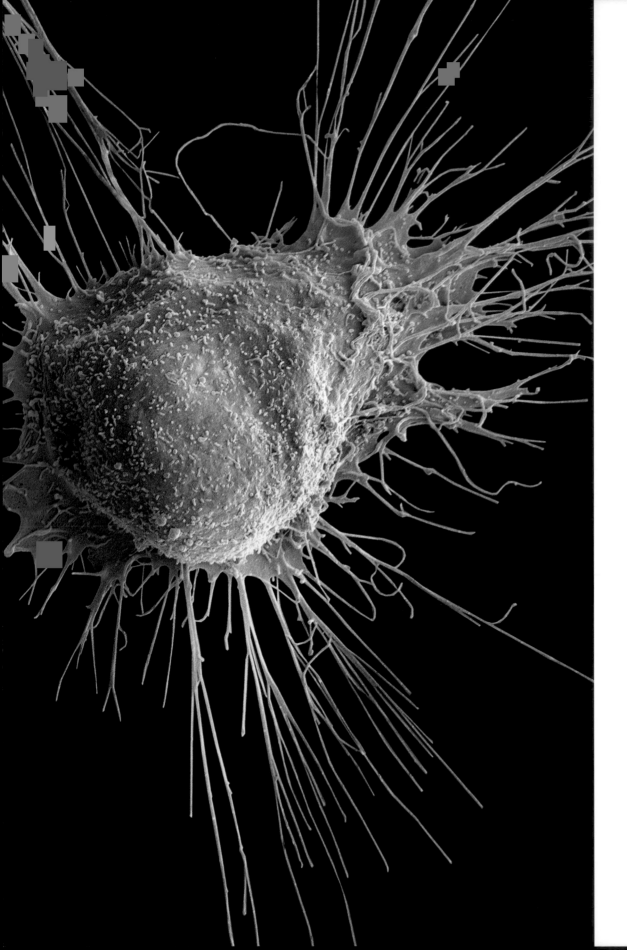

Left: Prostate cancer cell
(scanning electron micrograph)
The prostate is a gland just below the male
bladder, wrapped around the duct through
which the bladder empties its urine. Cancer
of the prostate, which is commonest in men
over 50, can restrict the flow of urine and
cause considerable discomfort. Mutations
in the DNA of cells are the general cause of
cancers, but the precise factors that trigger
prostate cancer are not known. Prostatic
tumours are slow-growing and usually
diagnosed at an early stage.
(Magnification x2000 at 10cm wide)

Right: Breast cancer cell
(scanning electron micrograph)
This highly magnified image of a single breast
cancer cell shows clearly its typically uneven
surface. Cancer cells are by their very nature
abnormal; they multiply rapidly, irregularly
and incompletely, which explains their ragged
appearance. They combine, as tumours, to
invade the surrounding tissue and can disperse
through the body causing secondary tumours.
It is the most common cancer among women
and is fully treatable by the surgical removal
of tumours combined with chemotherapy or
radiotherapy.
(Magnification unknown)

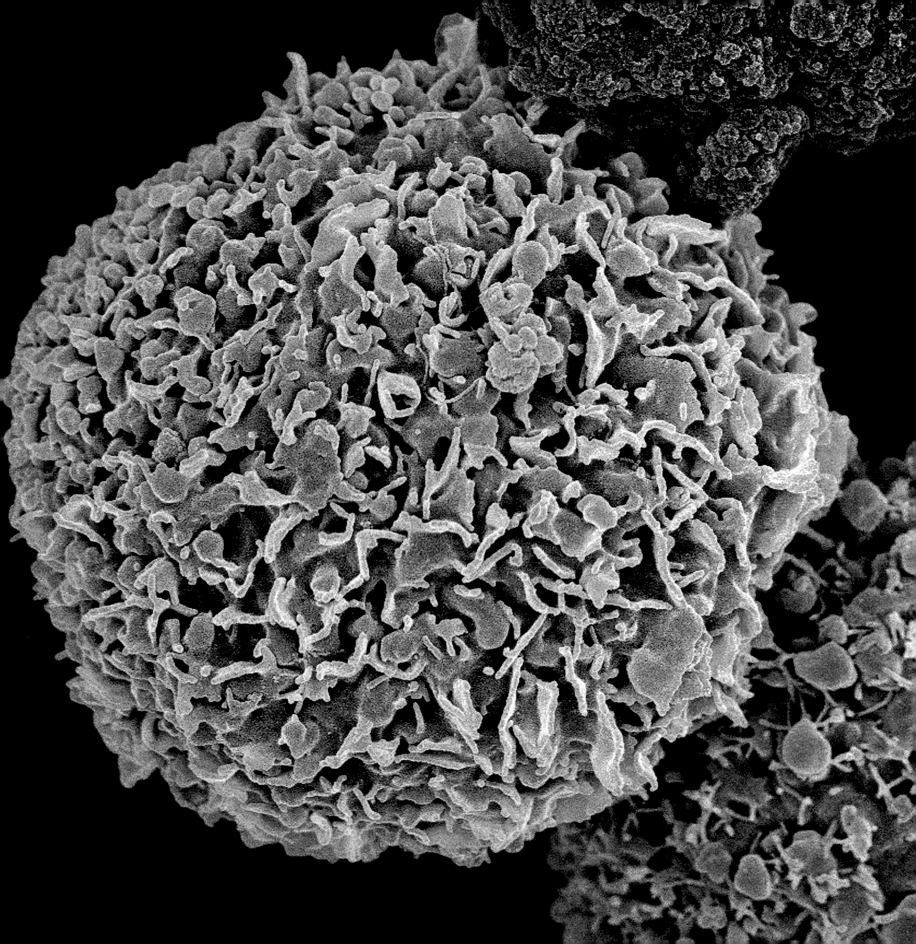

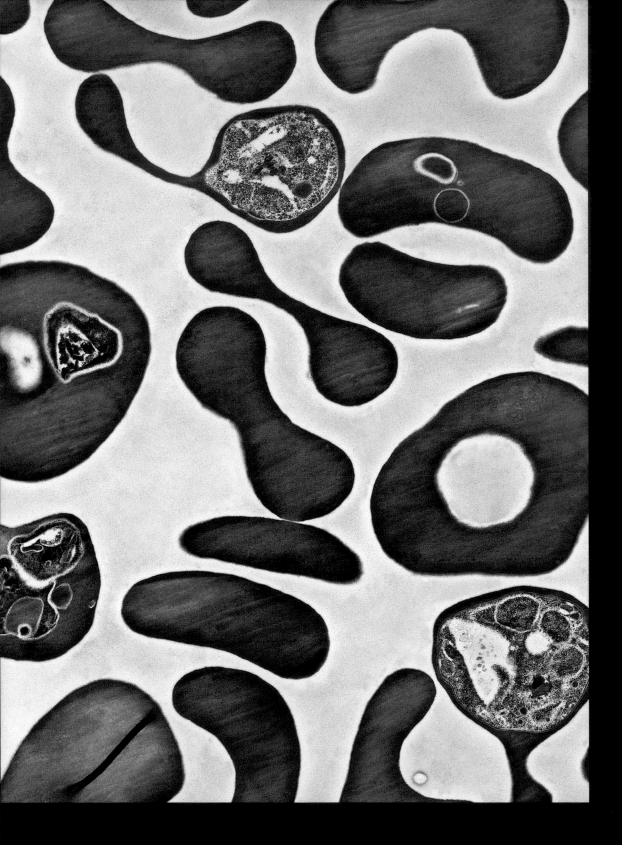

**Left: Malaria parasites
(transmission electron micrograph)**
In this image of red blood cells it's easy to see
which ones have been inhabited by malaria
parasites. The parasites occupy the cells and
consume the haemoglobin. They then replicate
themselves and colonize new cells in the search
for nourishment. For the infected human, this
loss of haemoglobin causes anae   le
each new wave of colonization triggers a bout
of malarial fever, every two or three days
depending on the strain of malaria involved.
(Magnification unknown)

**Right: Malaria-infected red blood cell
(scanning electron micrograph)**
The malaria parasite goes through several
stages in its life cycle. When it first enters the
body, following a bite from a mosquito, it is
called a sporozoite and attacks liver cells. There
it replicates as thousands of merozoites, which
burst out of the liver cells and infect red blood
cells – this image shows several healthy round
red blood cells and one distorted ne which
has been infected by a merozoit   There it will
replicate as twenty or so more merozoites,
bursting out again to continue the cycle.
(Magnification x7,000 at 10cm wide)

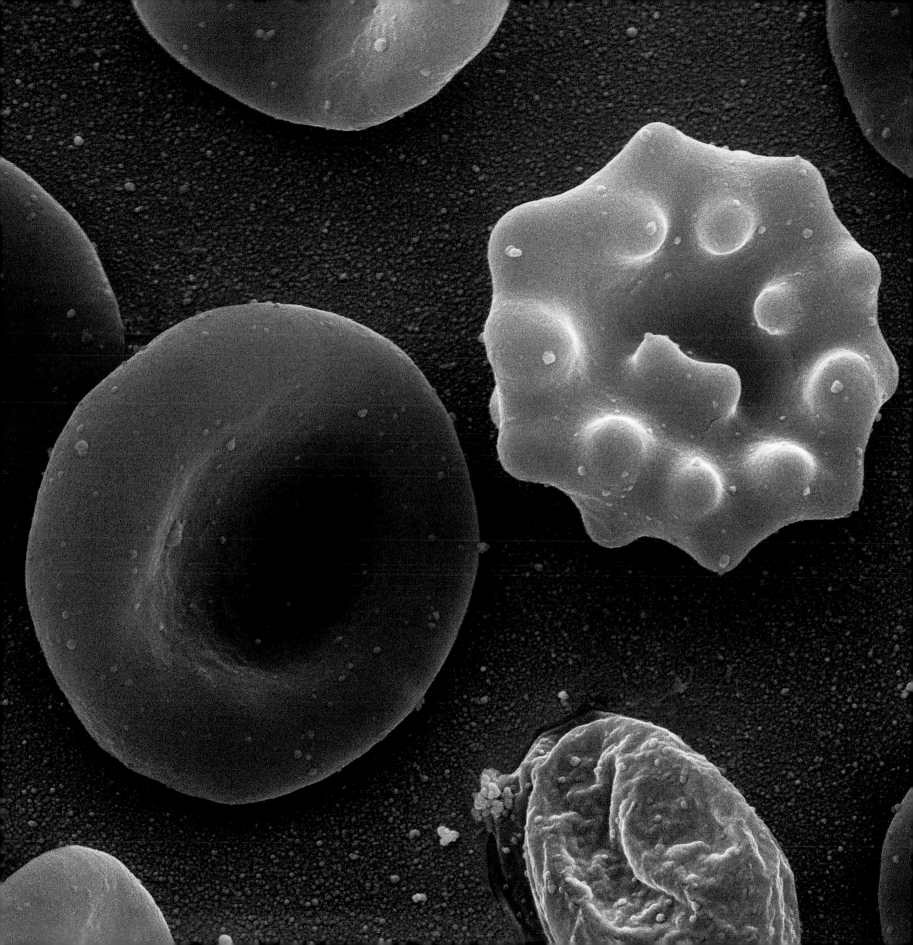

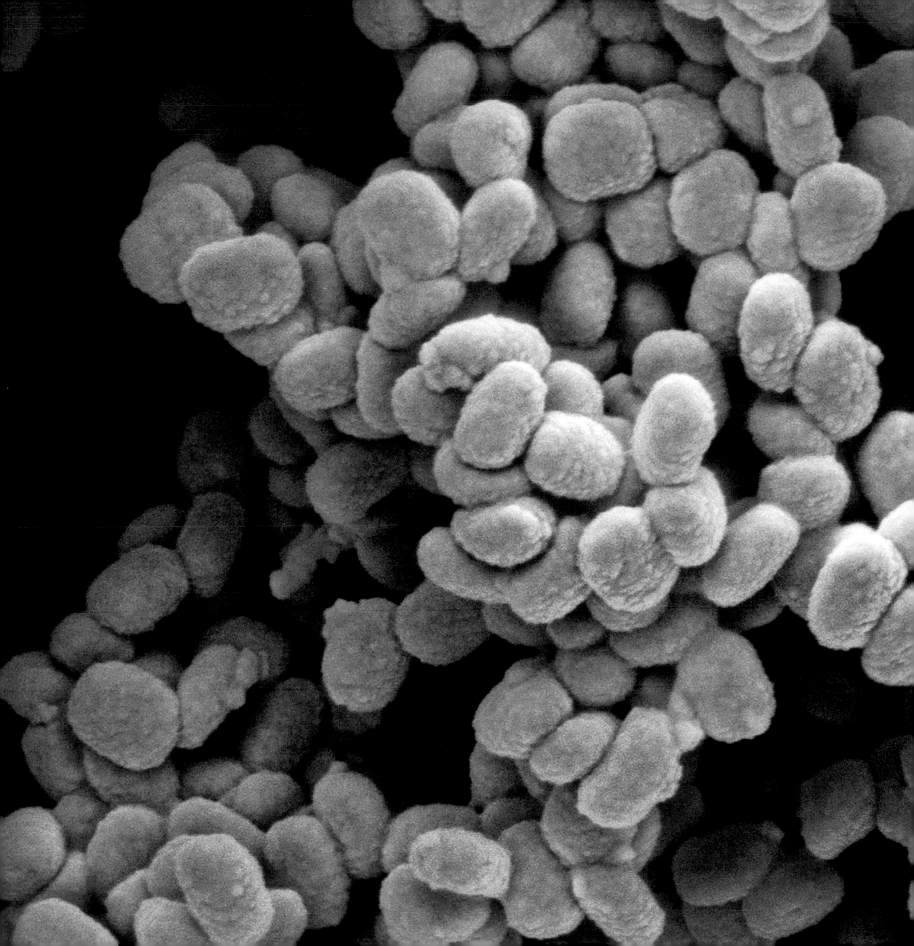

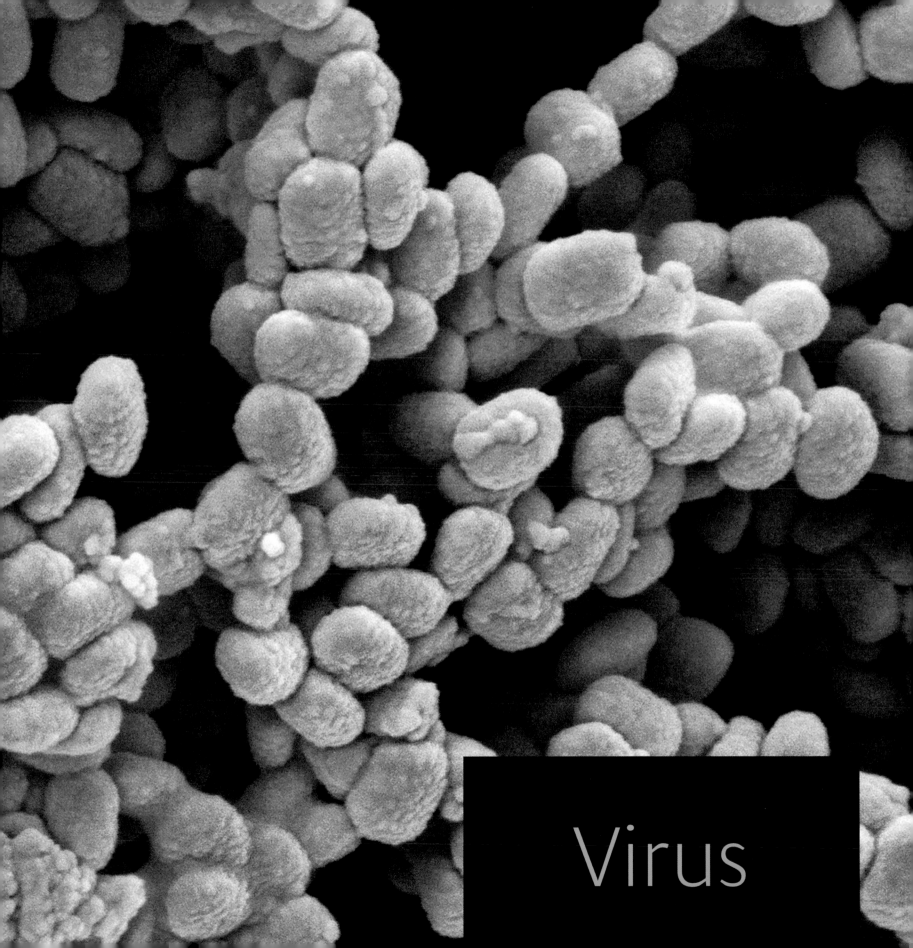

Virus

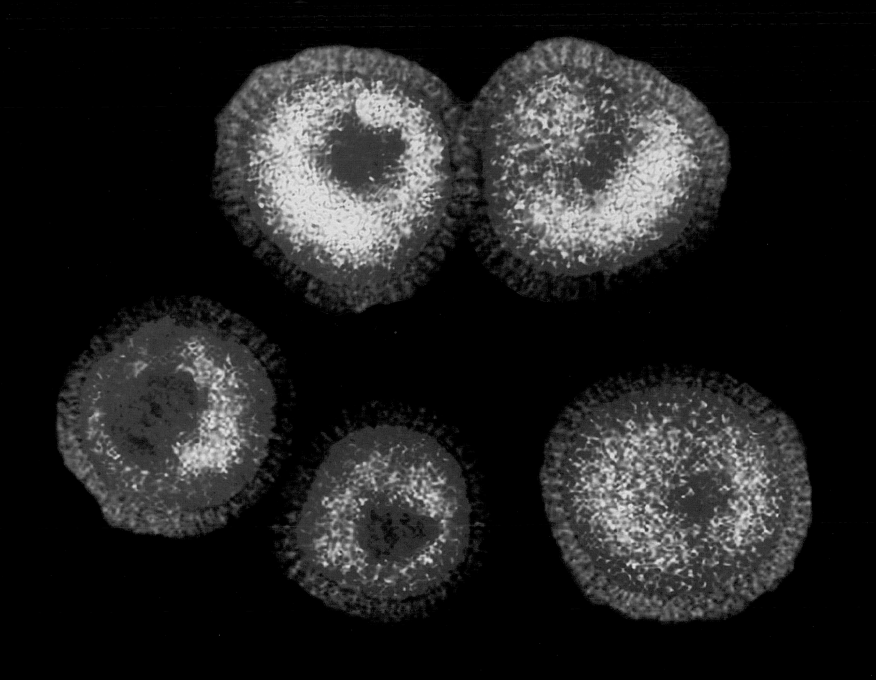

Previous page: Pox virus (scanning electron micrograph)

MCV, the *Molluscum contagiosum* virus, is (as its name suggests) highly contagious. Most people however are immune to it – only those with weakened immune systems such as children, the infirm and the sexually active tend to be affected. The resulting disease, known as the pox or water warts, is caught not only by contact with other infected people but from items which an infected person may have touched, including clothing and unwiped furniture.
(Magnification x20,000 at 10cm wide)

**Above: Influenza virus or Swine flu virus
(transmission electron micrograph)**

These flu virus particles, called virions, are of the strain known as swine flu, or H1N1 – a reference to its composition of the glycoproteins haemagglutinin (H) and neuraminidase (N). It was responsible for a serious flu epidemic in 2009, and probably also the so-called Spanish flu pandemic of 1918. The 1918 outbreak affected 500 million people worldwide, of whom between 50 and 100 million died – 3 to 5% of the world's population at the time. (Magnification unknown)

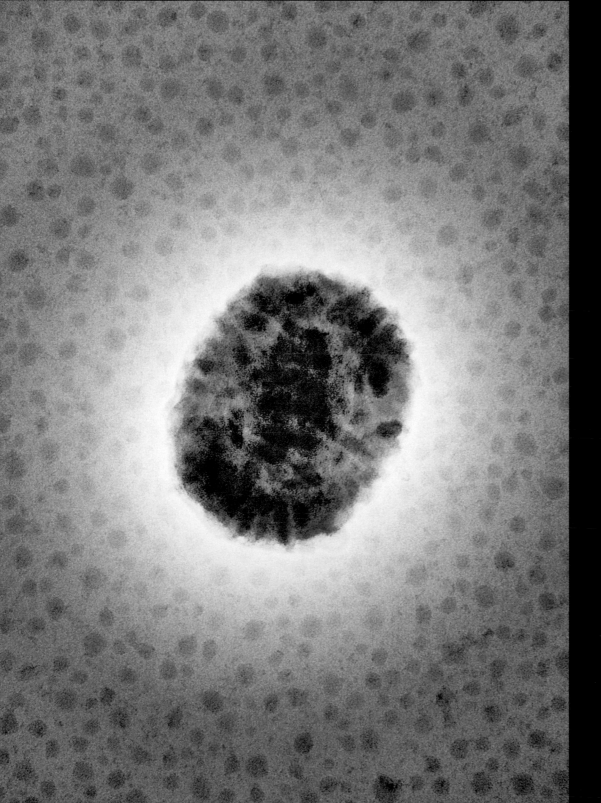

**_Molluscum contagiosum_ virus
(transmission electron micrograph)**
This is a single particle of the very contagious
pox virus. Those with weak immune systems,
the very young or very old, and sexually
active people are most susceptible to it: it
manifests itself as small round domed lesions
on the skin of the limbs, torso or groin that give
the disease its common name, water warts.
The warts clear up without treatment
or with over-the-counter medicines, usually
in a few months.
(Magnification 83,000 at 10cm tall)

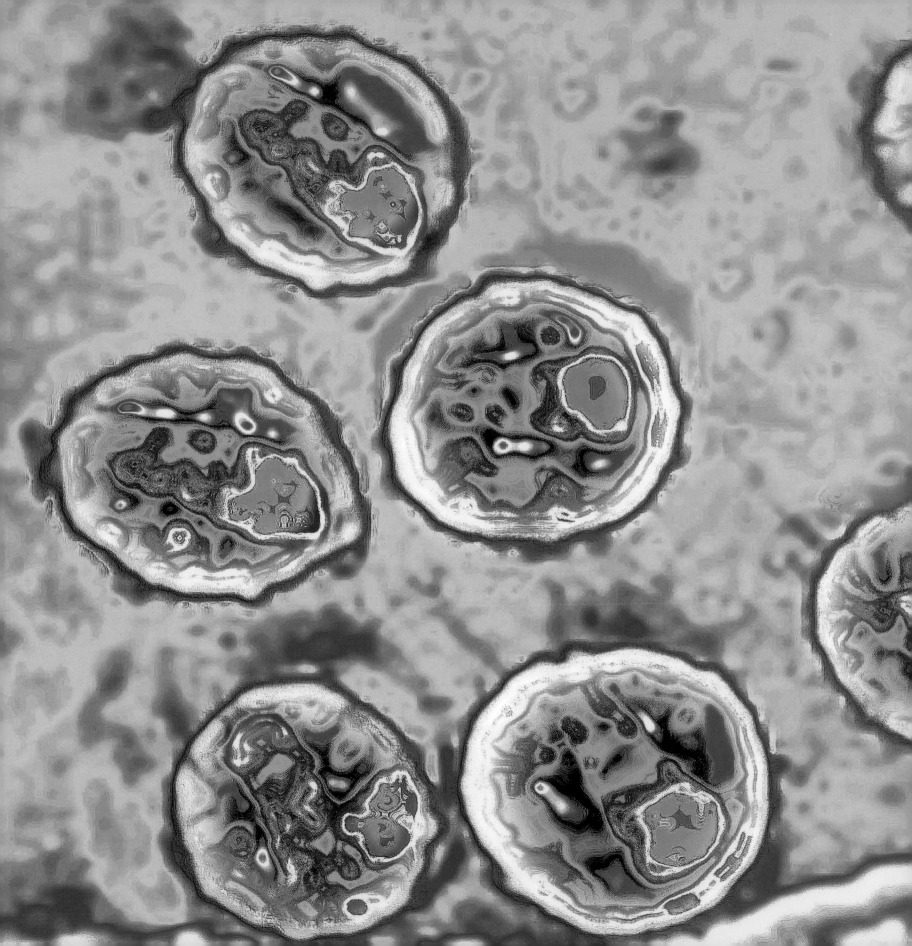

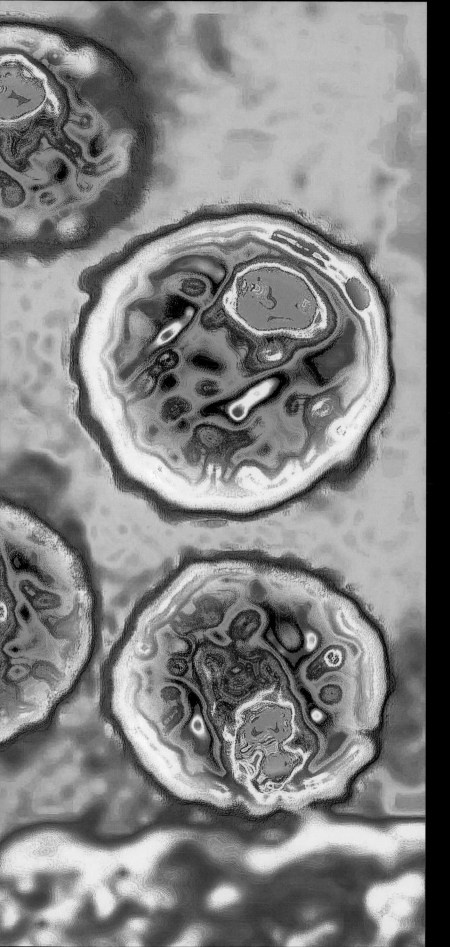

HIV (transmission electron micrograph)
This image shows HIV particles about to attack a white blood cell (the yellow edge at the bottom of the picture). The human immunodeficiency virus (HIV) invades these blood cells, called lymphocytes, replicates within them and then destroys them while the new HIV particles move on to infect more cells. Lymphocytes play a vital role in the human immune system, and their depletion makes HIV patients highly vulnerable to disease and infection.
(Magnification unknown)

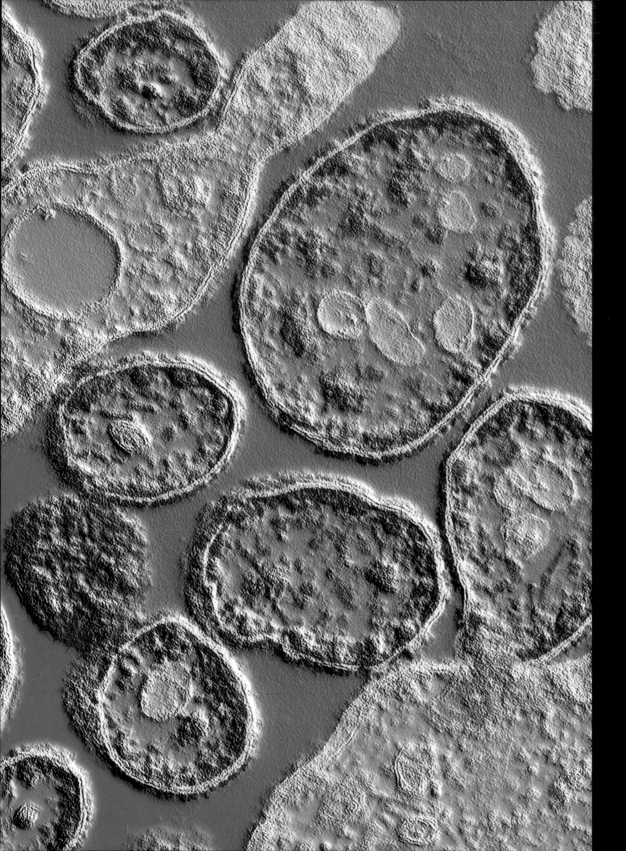

**Left: Measles virus infection
(transmission electron micrograph)**
Although measles virus relies entirely on
human hosts, it is believed to be a mutation of
an old virus affecting cattle, called rinderpest.
Rinderpest was eradicated in 2001; and in
2016 a programme of vaccination eliminated
measles from the Americas. In this image the
protein that the virus uses to bind to the host
cell is coloured purple. At bottom right, a
measles particle is breaking away from the host
cell (grey) to infect another elsewhere.
(Magnification x107,000 at 10cm tall)

**Right: Mumps virus
(transmission electron micrograph)**
In this image of a single mumps virus
particle (called a virion) the protein-coated
shell or capsule of the particle is shown in
pink. The linear features within it (in red) are
strands of ribonucleic acid (RNA). The RNA
interacts with the shell protein in a way that
allows the virus to connect to a host cell, infect
it, replicate within it and finally move on to
infect another host. Vaccination is effective in
preventing infection.
(Magnification unknown)

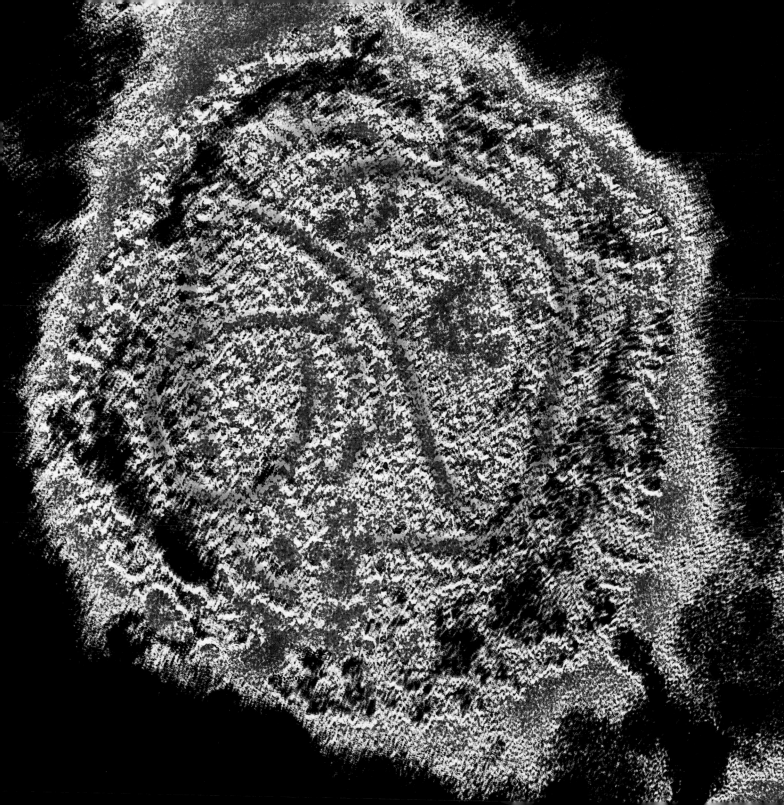

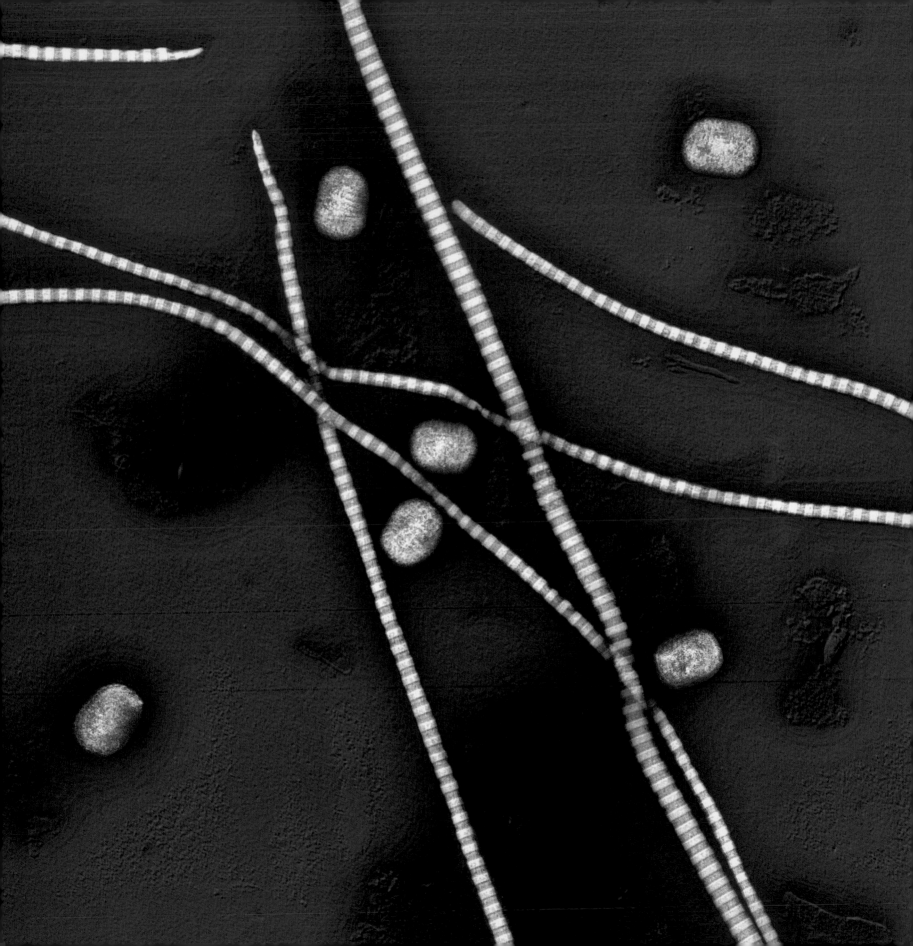

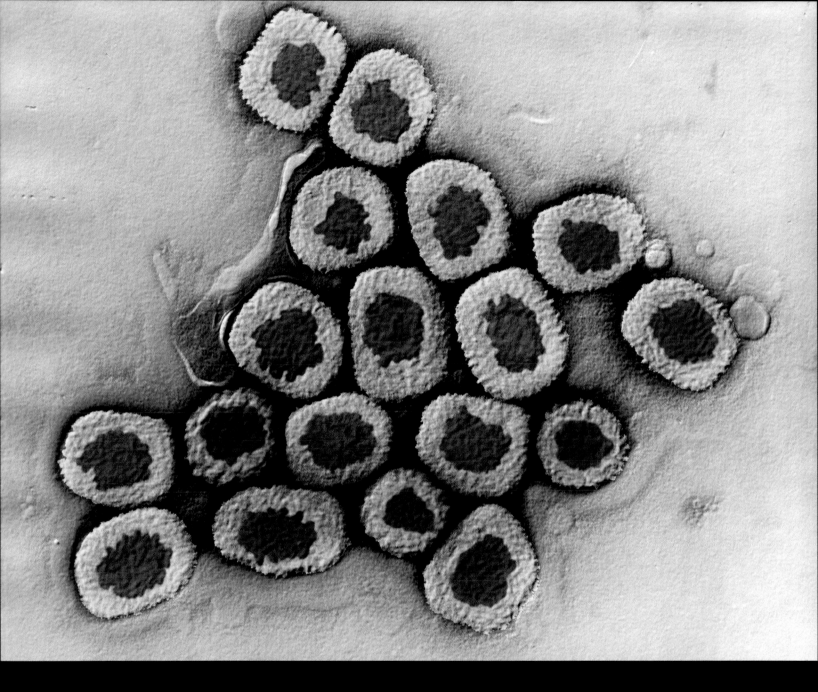

Left: *Molluscum contagiosum* virus (transmission electron micrograph)
The small oval forms (coloured orange in this image) are particles of the *Molluscum contagiosum* virus, a highly contagious infection that gives us the pox, also known as water warts because of the outward symptoms of the infection. The yellow strands in the picture are collagen fibres. Collagen is the material that fills the gaps in the body between cells, through which the particles must navigate as they seek new cells to infect.
(Magnification x25,000 at 10cm wide)

Above: Variola viruses (transmission electron micrograph)
Looking for all the world like a portion of sushi, this is in fact a cluster of variola viruses, which give us smallpox. Like many viruses it has a genetic core (here coloured red) surrounded by a protein casing (yellow), and it is the interaction between the two that allows the virus to latch on to a human cell and infect it. Variola now survives only in a few laboratories: smallpox was eradicated through vaccination in the 1970s.
(Magnification x63,000 at 10cm wide)

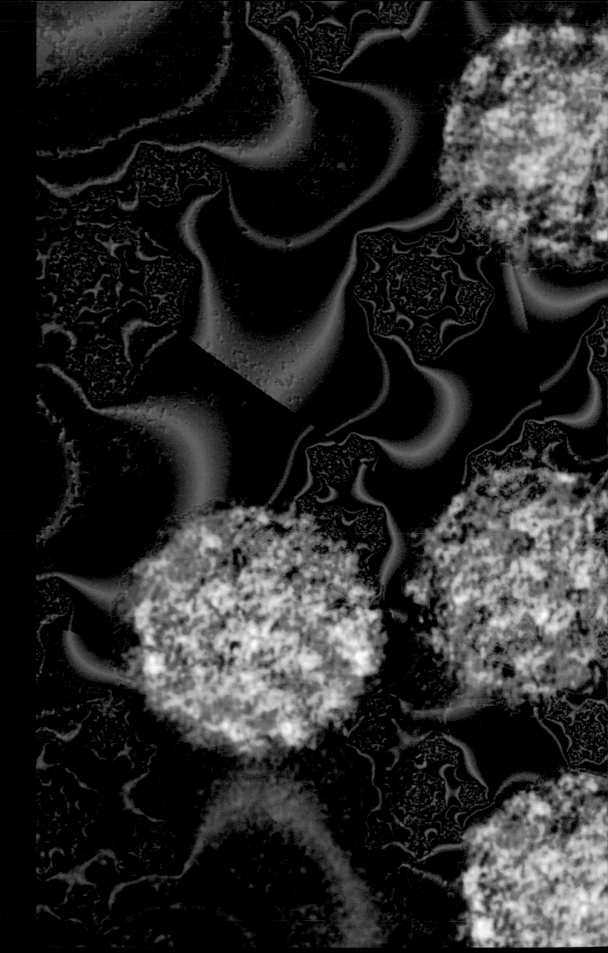

Papilloma viruses
(transmission electron micrograph)

The striking artificial colouring of these papilloma virus particles, and their superimposition on a fractal background, shouldn't distract you from their primary function, which is to give you warts. Human papilloma virus (HPV) is the most common sexually transmitted disease, and results in warts in and around the genitals, anus or throat. Like others in the papovavirus family, HPV is a DNA virus, and (as you can see here) has no enclosing protein capsule or shell. (Magnification unknown)

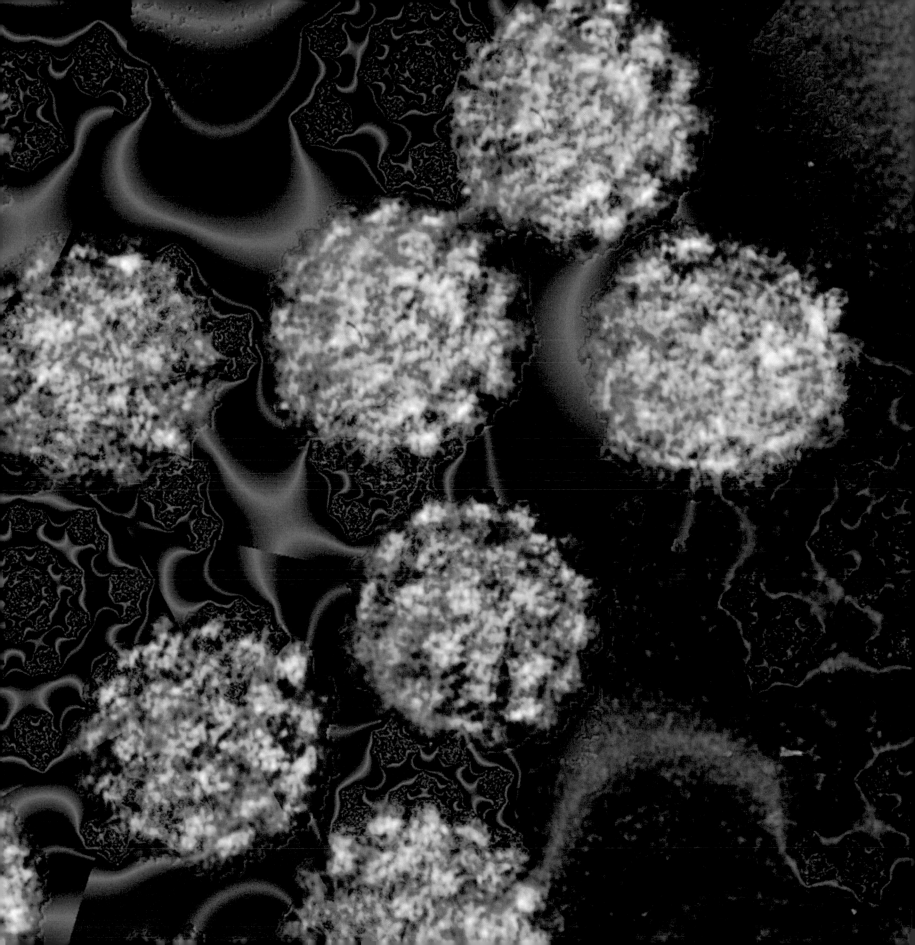

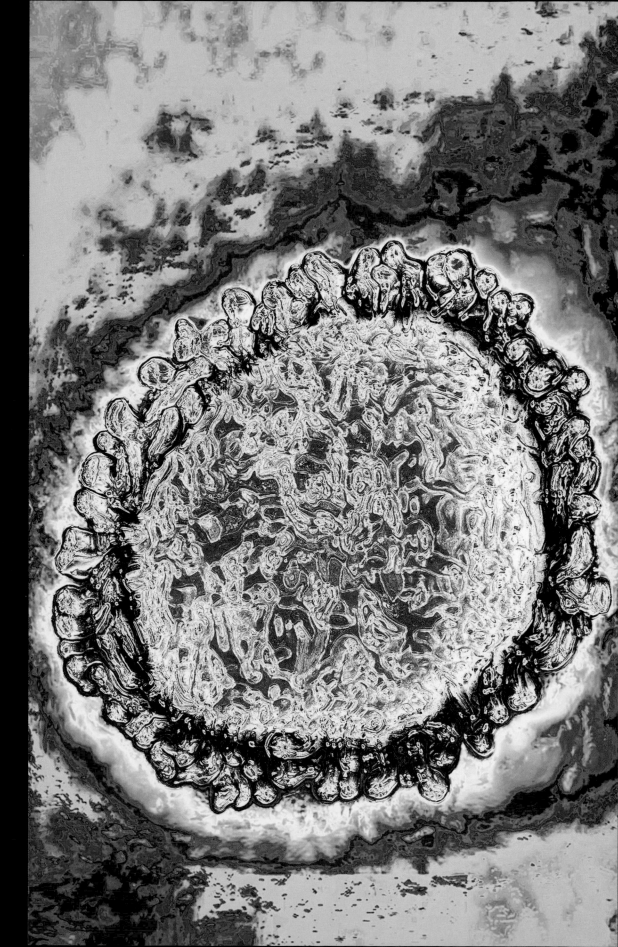

Left: Papilloma virus (HPV)
(transmission electron micrograph)
This is a single papilloma virus particle. There
are over 170 strains of human papilloma virus
(HPV). They cause warts on hands and feet,
and in several tracts of the body including the
throat, genitals and anus. Although the warts
themselves are considered non-malignant
tumours, some strains can increase the risk of
cancer in those areas. Vaccination at a young
age, before sexual activity has begun, is an
effective prevention.
(Magnification x2,000,000 at 10cm wide)

Right: Coronavirus
(transmission electron micrograph)
The coronavirus gets its name from the Latin
word for a crown, a reference to the fringe of
protein buds which ring the virus particles.
Most coronaviruses only give us heavy colds
and sore throats, but if they spread to the lungs
they can cause pneumonia. The same family
is also responsible for Middle East respiratory
syndrome (MERS) and severe acute respiratory
syndrome (SARS), which caused many deaths in
the early twenty-first century.
(Magnification x830,000 at 10cm wide)

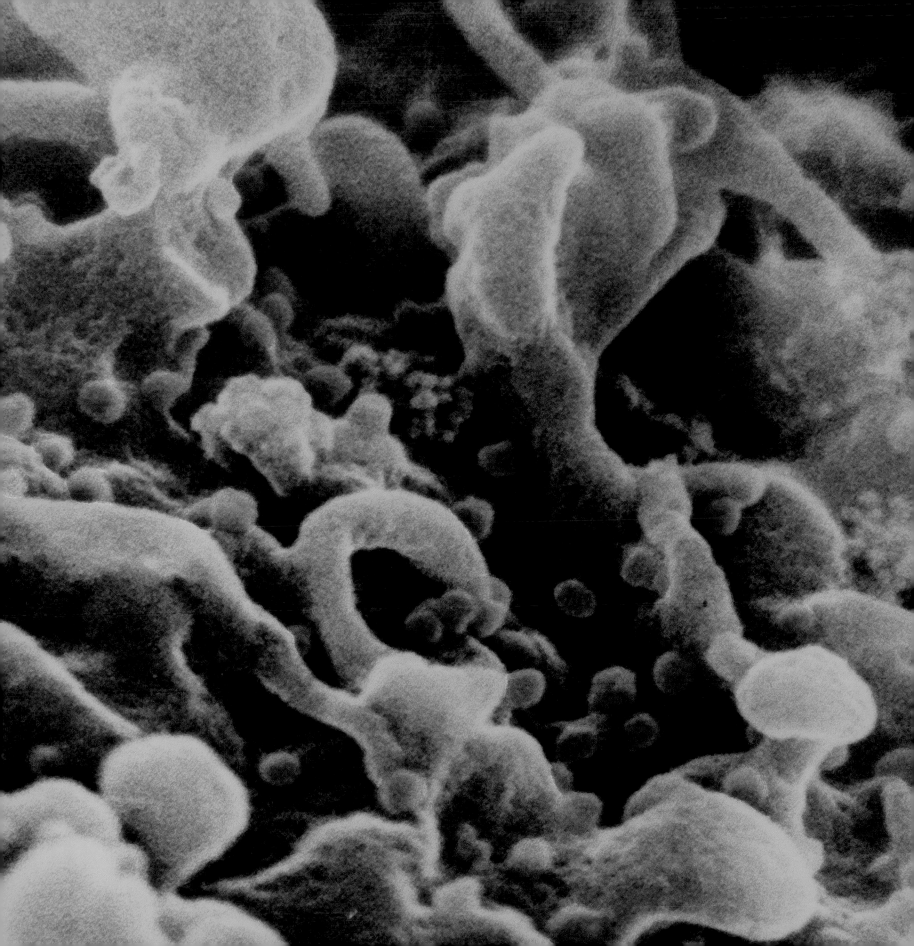

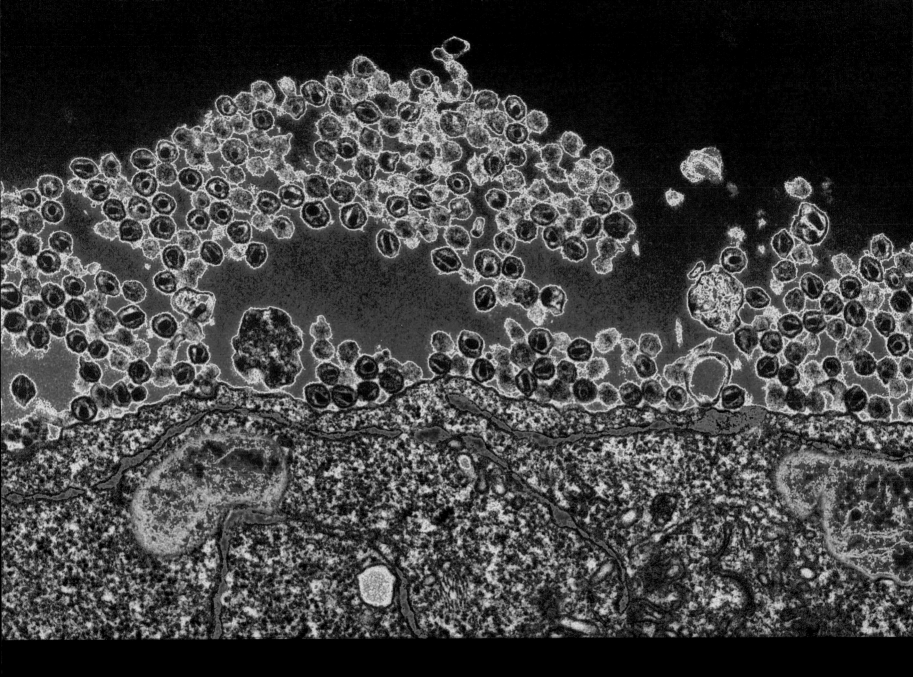

Left: HIV virus (coloured scanning electron micrograph)
This Christmassy scene of holly and berries is anything but festive. The red
particles are virions of the human immunodeficiency virus (HIV), attacking
the uneven green surface of a white blood cell. Such invasions eventually kill
off infected white cells, which are essential in defending the body against
disease. After 10–15 years of HIV infection, the body's immune system is
so damaged that acquired immune deficiency syndrome (AIDS) develops,
leaving patients extremely vulnerable to normally harmless infections.
(Magnification x51,300 at 10cm tall)

Above: HIV viruses budding from a T-lymphocyte blood cell (T-cell)
(transmission electron micrograph)
The grainy pink mass at the bottom of this picture is a white blood cell,
which has been infected by HIV. The infecting virus persuades the cell to
produce more HIV particles, which then burst out of the cell to invade
other cells, leaving the host cell depleted and dying. This image captures
the moment at which the replicant HIV particles (seen as small round
purple bubbles) depart from the doomed host cell.
(Magnification x10,850 at 7cm wide)

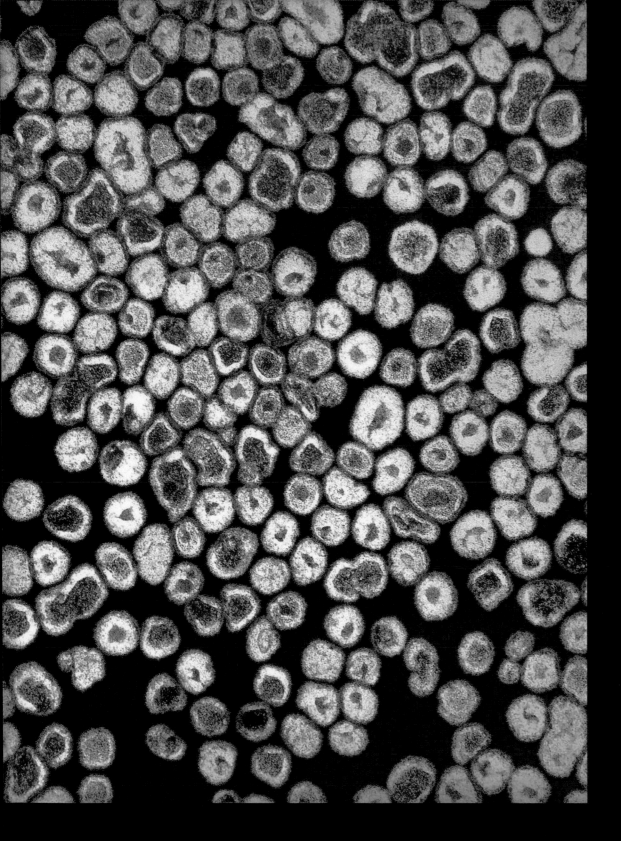

The La Crosse virus was first identified in La Crosse, Wisconsin, USA. Humans may be infected by the bite from a forest mosquito, or from a forest animal (for example a squirrel or chipmunk) which has been bitten by a mosquito. The virus causes La Crosse encephalitis, an inflammation of the brain which results in fever and nausea. In rare, extreme cases, patients can suffer seizures, comas and even permanent brain damage.
(Magnification unknown)

Right: Bacteriophages attacking bacteria
(transmission electron micrograph)
A bacteriophage is a virus that attacks bacteria rather than cells. Typically it consists of a 20-sided head and a tail, with so-called tail-fibres which fix it to the host bacterium. It injects its DNA into the host, which then replicates the bacteriophage before releasing the replicants to attack further bacteria. Since the early twentieth century bacteriophages have been used as an alternative to antibiotics for the treatment of certain drug-resistant strains of bacteria.
(Magnification unknown)

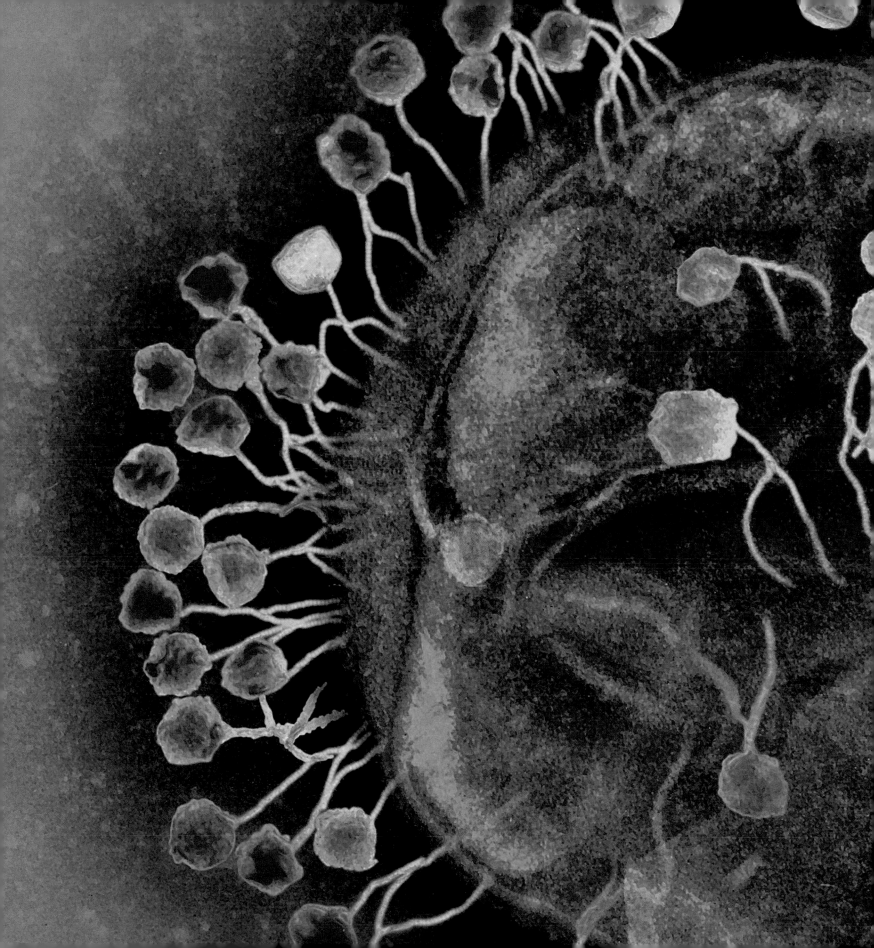

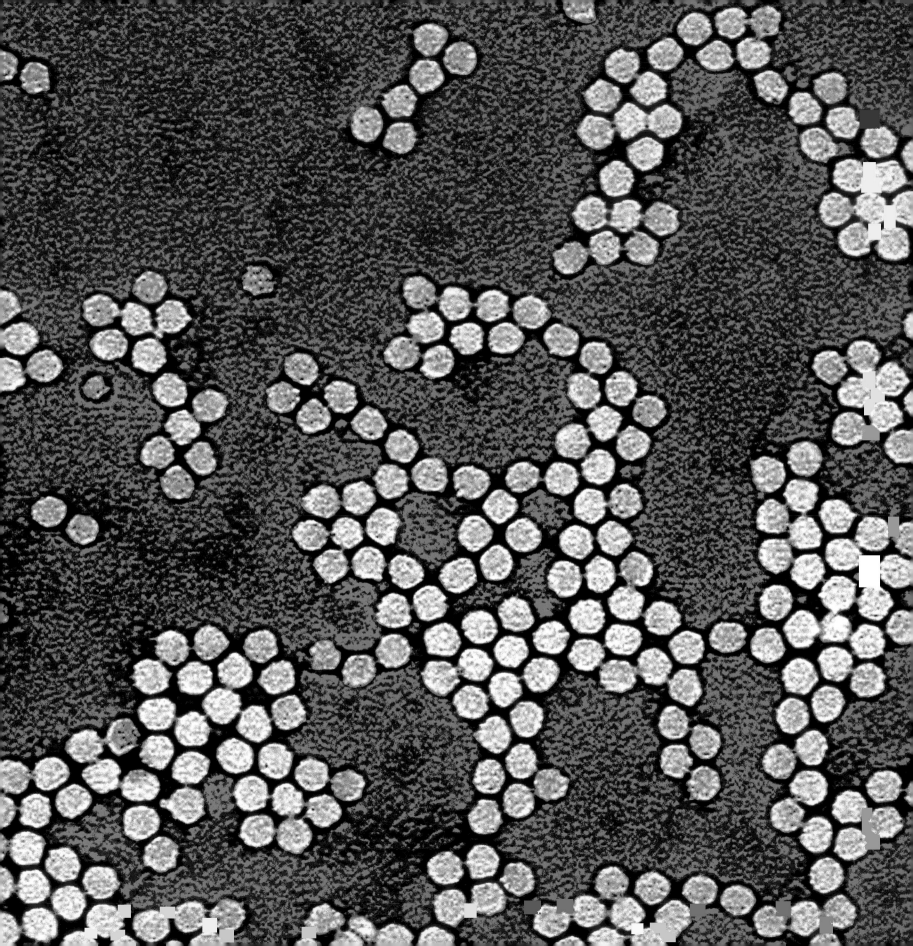

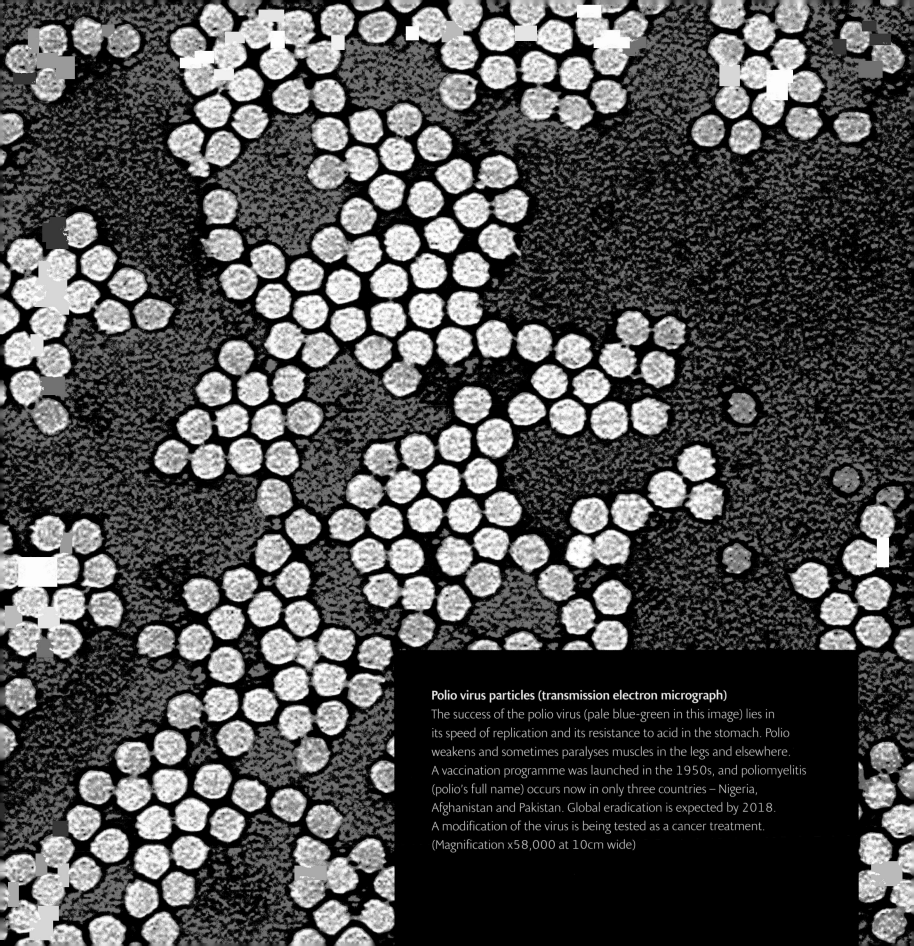

Polio virus particles (transmission electron micrograph)
The success of the polio virus (pale blue-green in this image) lies in
its speed of replication and its resistance to acid in the stomach. Polio
weakens and sometimes paralyses muscles in the legs and elsewhere.
A vaccination programme was launched in the 1950s, and poliomyelitis
(polio's full name) occurs now in only three countries – Nigeria,
Afghanistan and Pakistan. Global eradication is expected by 2018.
A modification of the virus is being tested as a cancer treatment.
(Magnification x58,000 at 10cm wide)

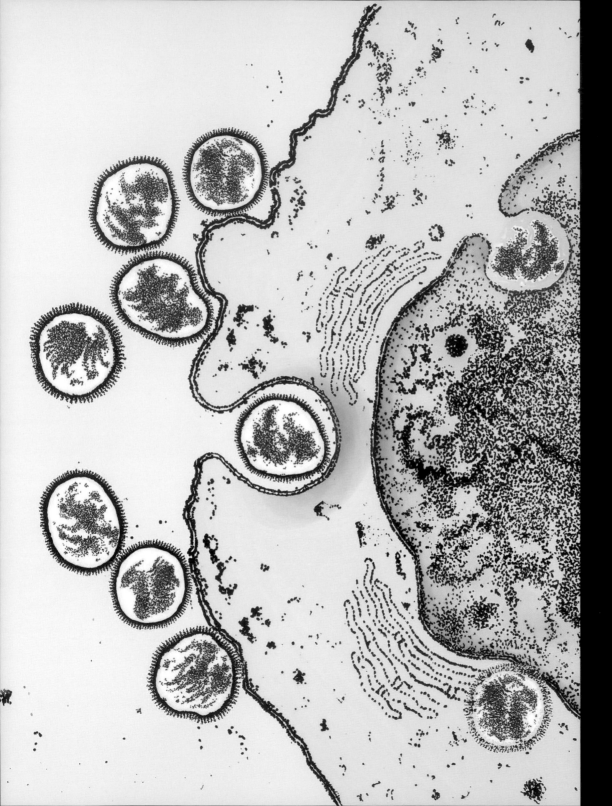

**Left: Avian flu
(transmission electron micrograph)**
Avian or bird flu is so called because it is adapted
to live in birds; but it can adapt infinitely into
strains against which humans have little or
no resistance. It weakens the lungs, causing
breathing problems and exposing them to
bacterial infections. The protein shell of the virus
encourages cells in the surface of the lung (here
yellow) to engulf it. Once inside, it sheds its shell
and the enclosed RNA attacks the cell nucleus
(shown here in green).
(Magnification unknown)

**Right: H5N1 avian influenza virus particles
(transmission electron micrograph)**
Outbreaks of avian flu have increased in frequency
since the 1990s. One of the most notorious is
the H5N1 strain, carried by immune wild birds,
which infect domestic poultry. Humans inhale
airborne virus particles (shown here in orange)
from poultry faeces, which infect cells in the lining
of the lung. H5N1 virus has killed over half of all
confirmed human cases (around 400 deaths)
since it was first recorded in 2003.
(Magnification x230,000 at 10cm tall)

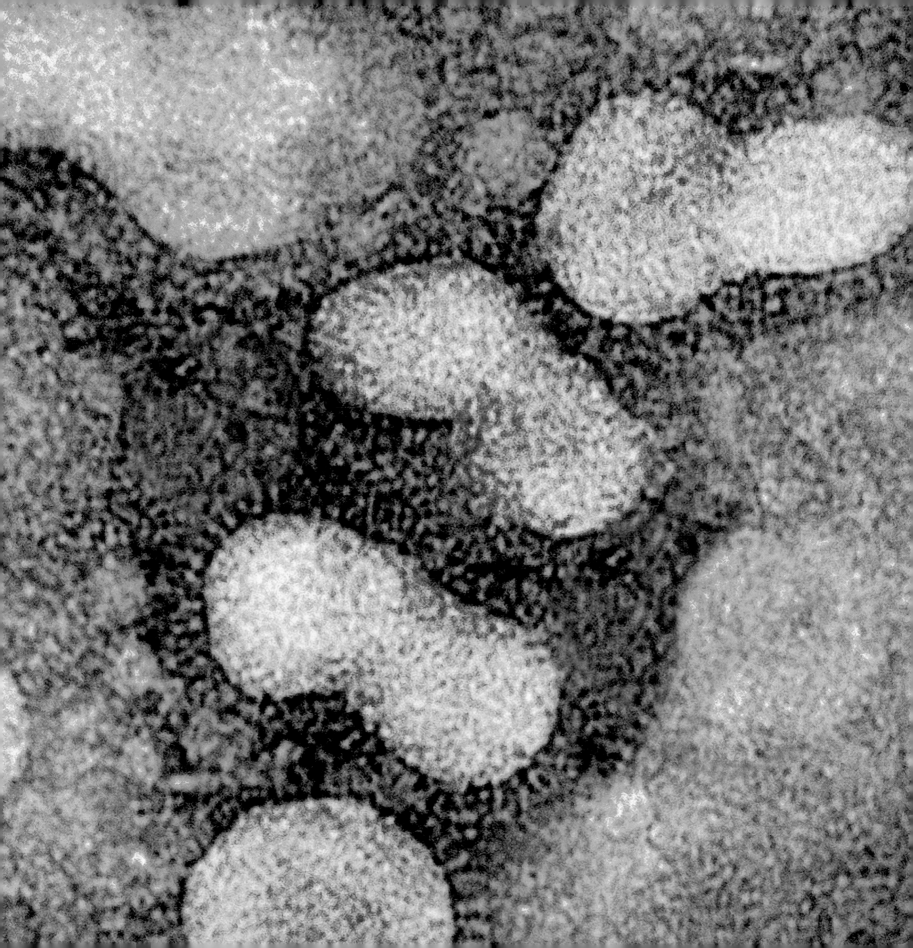

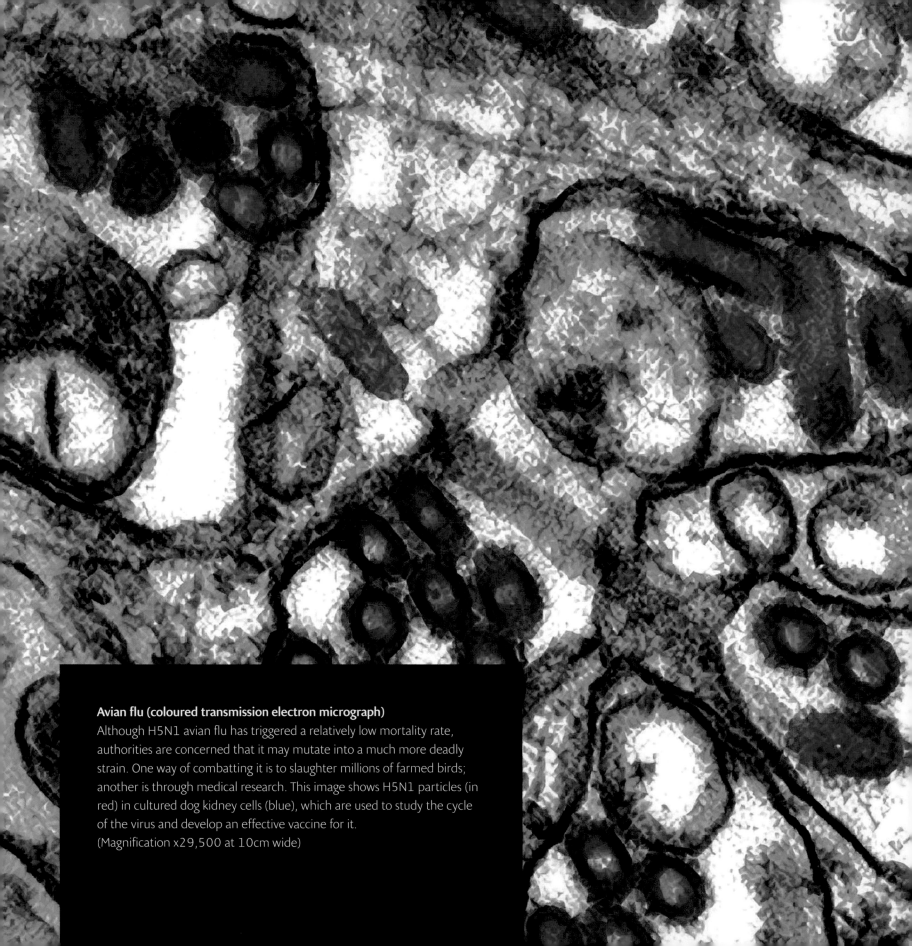

Avian flu (coloured transmission electron micrograph)
Although H5N1 avian flu has triggered a relatively low mortality rate,
authorities are concerned that it may mutate into a much more deadly
strain. One way of combatting it is to slaughter millions of farmed birds;
another is through medical research. This image shows H5N1 particles (in
red) in cultured dog kidney cells (blue), which are used to study the cycle
of the virus and develop an effective vaccine for it.
(Magnification x29,500 at 10cm wide)

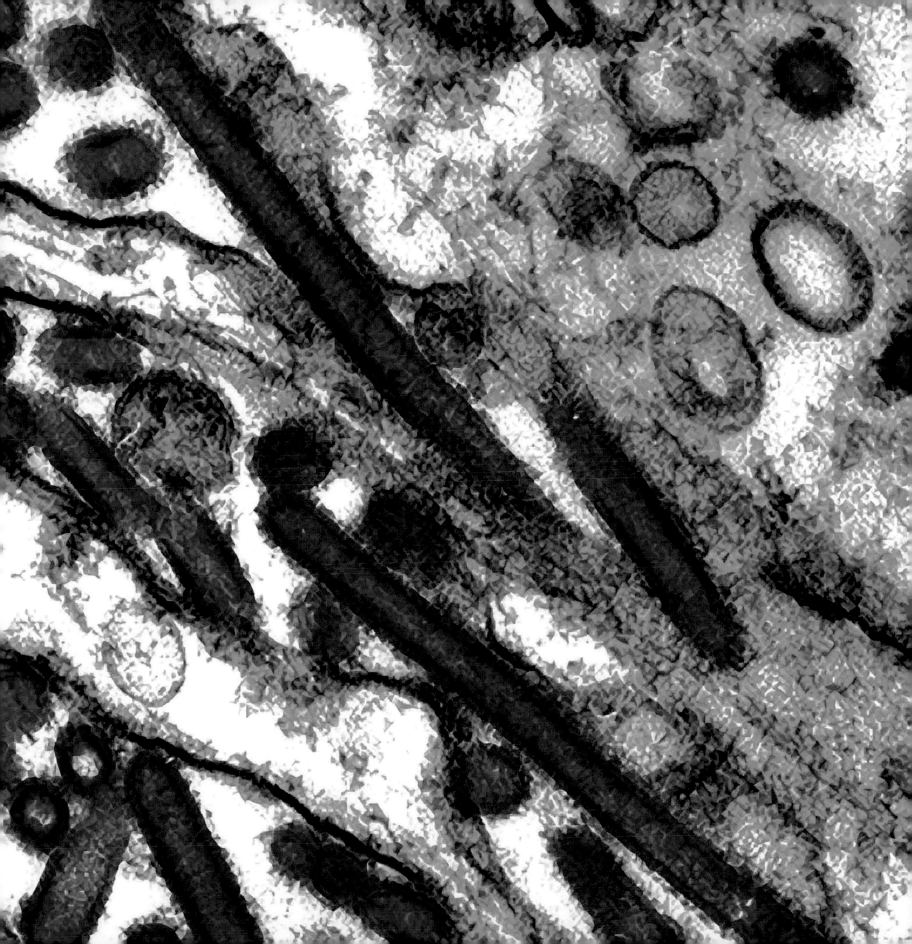

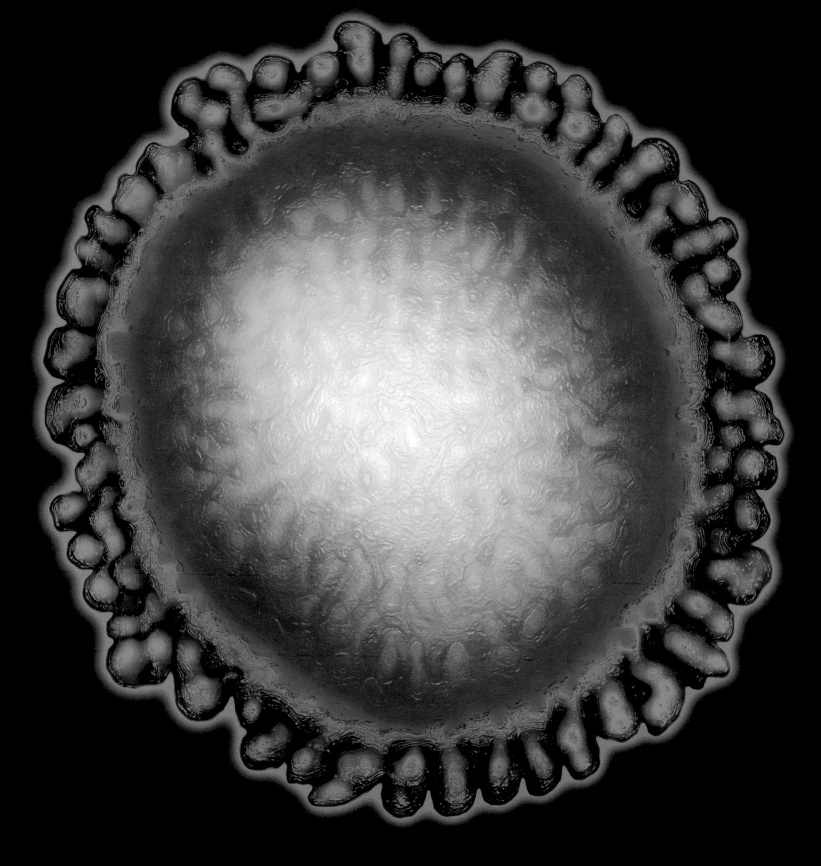

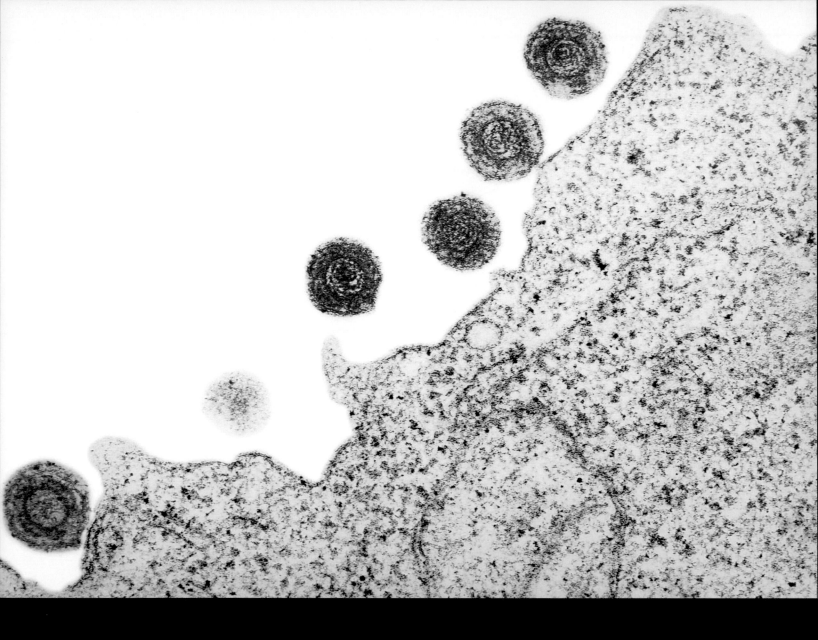

Left: Human coronavirus (transmission electron micrograph)
This image shows perfectly how the coronavirus got its name – from the Latin word for crown. The spikes (called peplomers) of the crown are made of proteins that help the virus decide which cells to attack. A peplomer needs to find the right receptor (a protein molecule) on a cell in order to fix on to it. Coronaviruses were first identified in 1960s, in the nose of someone with the common cold.
 (Magnification x1,000,000 at 10cm wide)

Above: Human herpes virus-6 (transmission electron micrograph)
There are nine human herpes viruses (HHVs). HHV6 was first discovered in 1986 in the blood of patients with AIDS. Subsequently a distinction was drawn between HHV6A, which is connected with multiple sclerosis and other neuroinflammatory diseases, and HHV6B, which (like HHV7) causes the childhood illness roseola. In this image, HHV6 particles (in red), which have been replicated in an infected white blood cell (green) are being released from it to infect further cells.
(Magnification unknown)

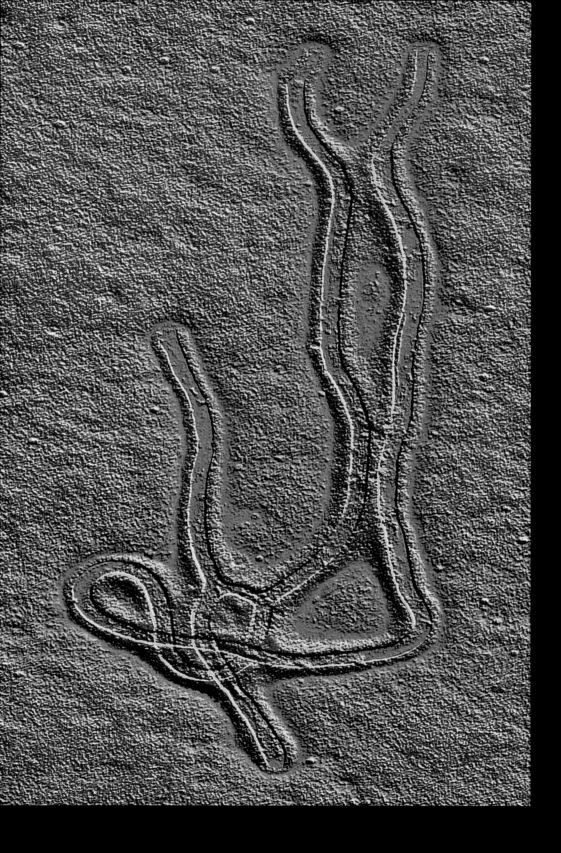

Left: Ebola virus (coloured transmission electron micrograph)

Ebola is a particularly unpleasant disease. An initial fever is followed by diarrhoea and vomiting, kidney and liver failure, and finally internal and external bleeding. Death occurs in around 50 per cent of cases, and the 2013–2015 outbreak in western Africa resulted in over 11,000 fatalities. The particles of ebola virus are simple strands of RNA, clearly seen here. They are transmitted by bodily fluids and therefore easily caught from those exhibiting ebola's symptoms.
(Magnification unknown)

Right: Papilloma viruses (coloured transmission electron micrograph)

Human papilloma virus infection (HPV) often passes without symptoms. But persistent cases may result in warts and lesions which greatly increase the chance of cancer developing. HPV accounts for nearly all cases of cervical cancer, for example. Of more than 170 variants of the papilloma virus, 40 are transmitted sexually. Vaccines are effective only if given before infection occurs, and many countries have introduced a programme of vaccination for girls under 15 years of age.
(Magnification x100,000 at 3.5mm wide)

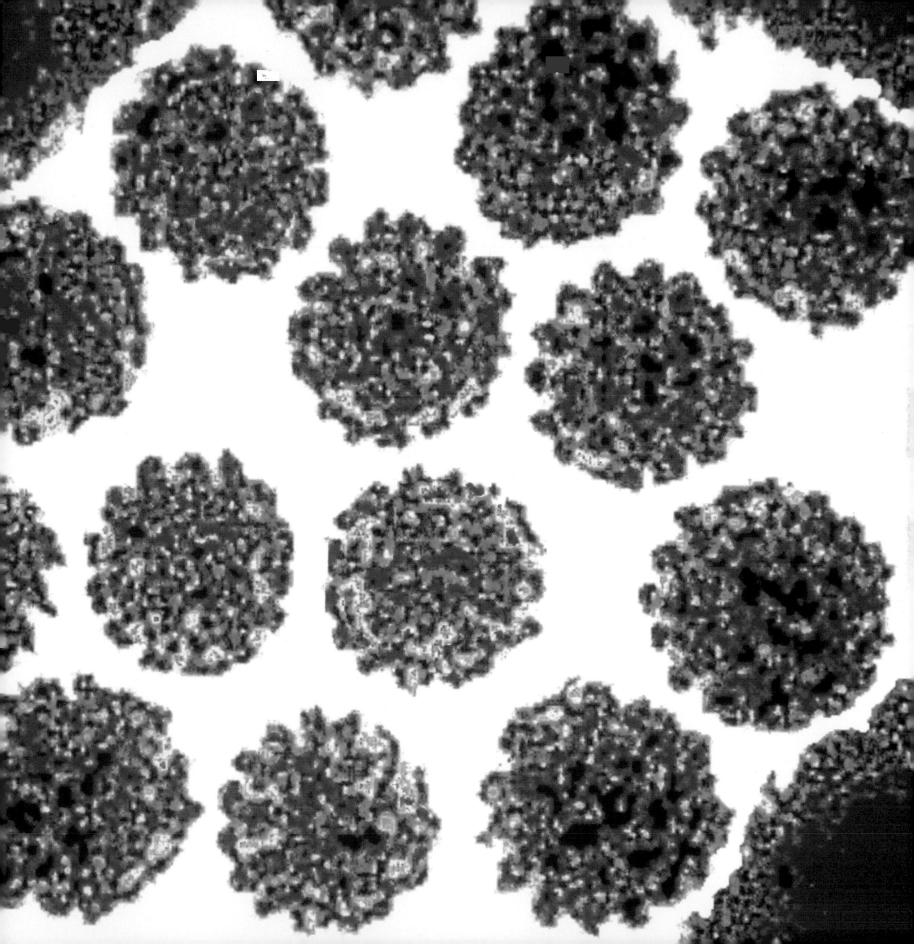

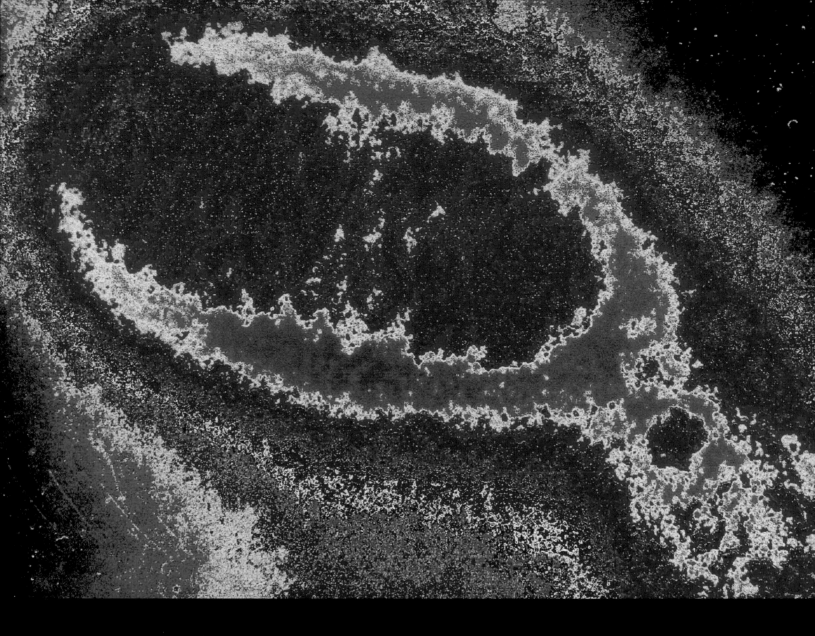

Above: Smallpox virus (transmission electron micrograph)
One of medical science's greatest achievements has been the global
eradication of smallpox, by a worldwide programme of inoculation.
Smallpox is caused by the variola virus (pictured here), which now survives
only as biological samples stored in a few secure laboratories in Russia
and America. It spreads via droplets of water in the breath, or from fluids
escaping from the disfiguring blisters that cover the bodies of its victims.
(Magnification unknown)

Right: Rotavirus (transmission electron micrograph)
Electron micrograph images like this are not just striking pictures. The very
existence of the rotavirus was discovered using an electron microscope,
in 1973. It's responsible for many childhood cases of diarrhoea, often
accompanied by vomiting and fever. It's vital to drink lots of water to
combat the severe and potentially fatal dehydration that follows. A
preventative vaccine is available, which has greatly reduced the frequency
and severity of infections.
(Magnification unknown)

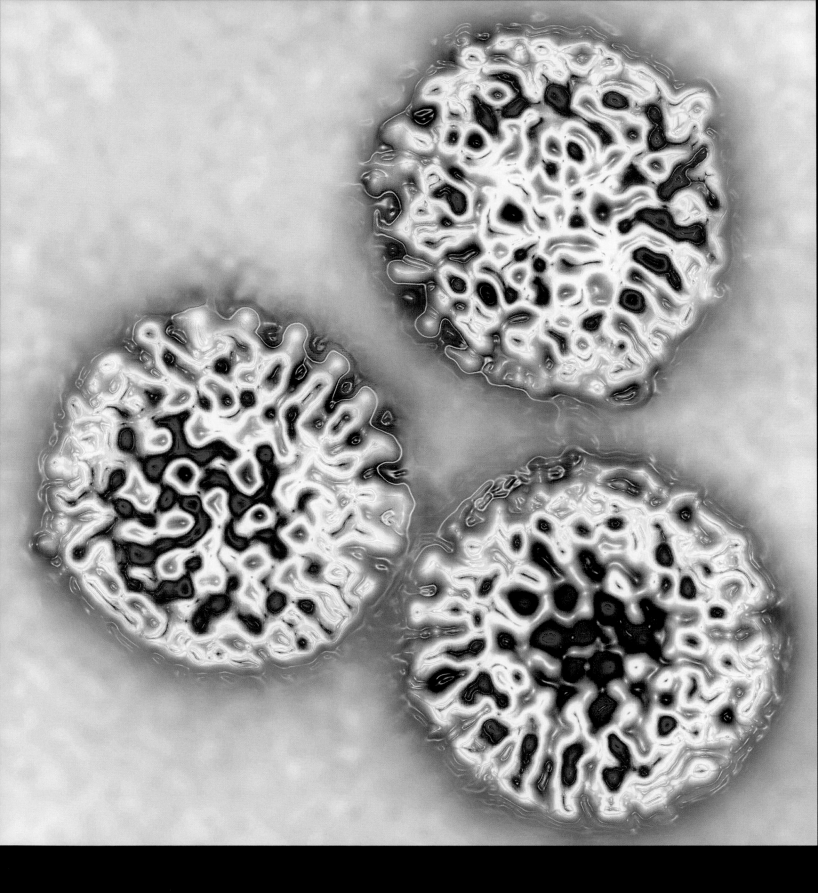

Virus 71

Above: Cell infected with HIV (transmission electron micrograph)
This image captures HIV particles (shown in pink) as they are about to
burst out of an infected cell (dark blue, at bottom). HIV attacks specialized
white blood cells call lymphocytes, upon which the body's defence
mechanisms rely. Since the lymphocytes die after infection, HIV eventually
destroys the immune system. Treatment for HIV usually involves a cocktail
of antiretroviral drugs, which can prevent its progression towards clinical
immunodeficiency in the form of AIDS.
(Magnification x90,000 at 10cm wide)

Right: HIV particles (transmission electron micrograph)
Although there remains as yet no effective HIV vaccine, drug and diet
therapies now make it possible for those acquiring the disease to live
with it as a chronic condition rather than to die of it by developing AIDS.
Infection occurs by one of three main routes – sexual contact; mother-to-
child, in the womb or by breastfeeding; and blood contact, for example
by sharing needles. We now know that kissing, for example, and sitting on
toilet seats, are not HIV risks.
(Magnification x210,000 at 20cm wide)

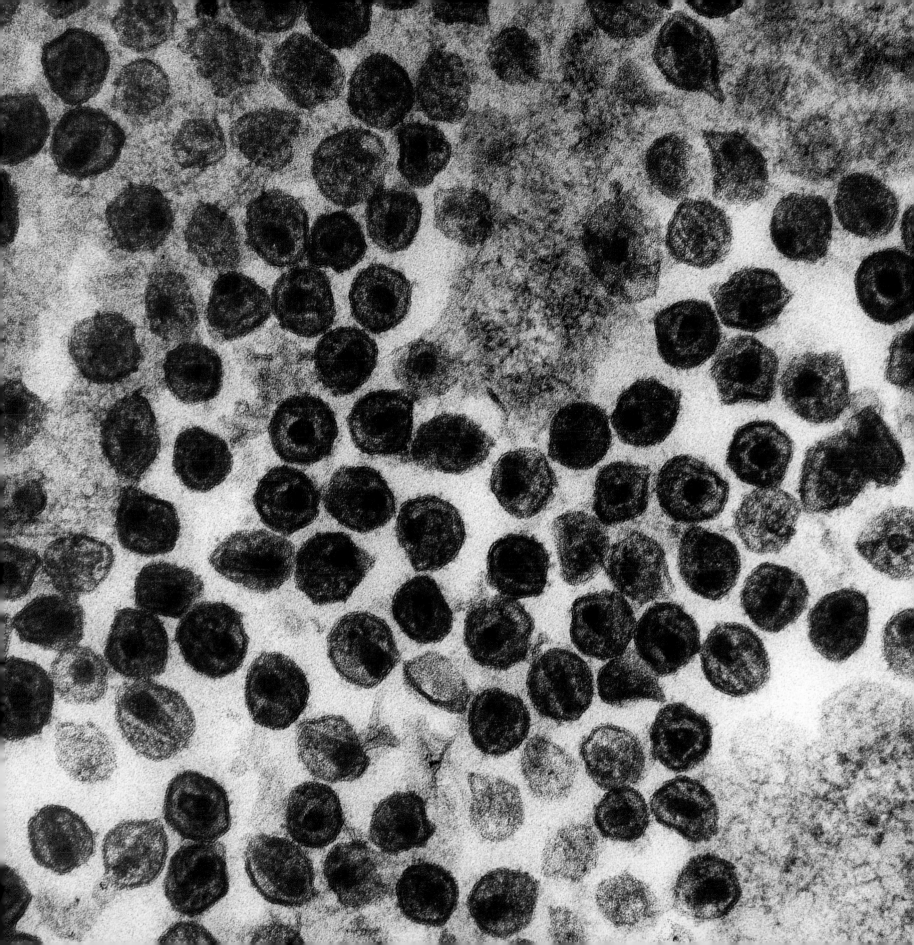

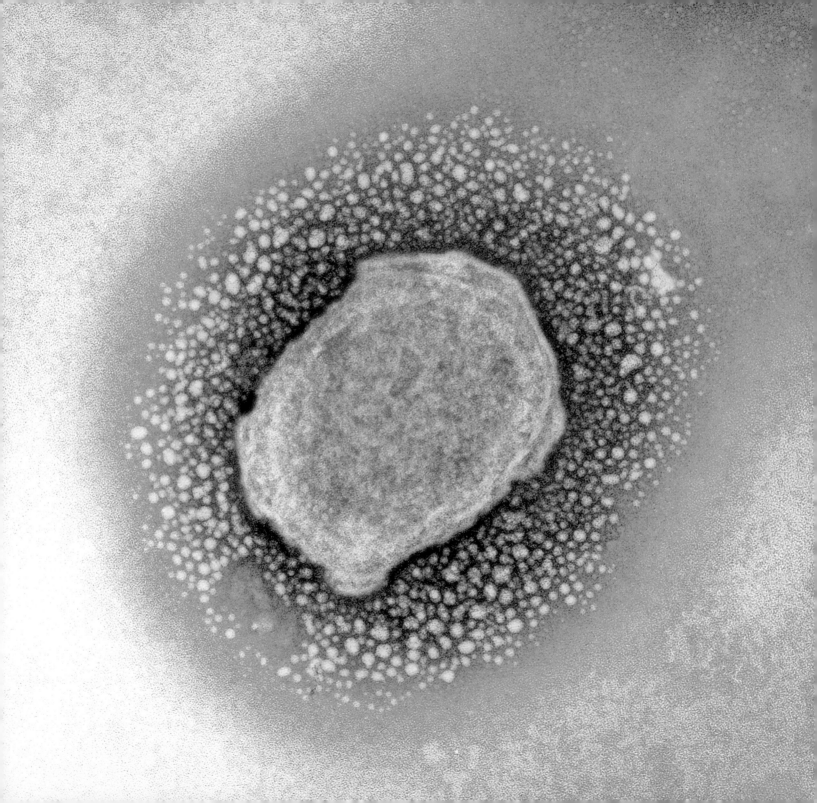

Left: Monkeypox virus particle (transmission electron micrograph)

Monkeypox is a milder relative of smallpox, although it can still be lethal. The symptoms are similar – a profusion of lesions over the body. It was first identified in 1958 in macaque monkeys, and first found in humans in 1970. Transmission is by bite or by contact with infected body fluid. The smallpox vaccine was considered good against both diseases, but since smallpox has been eradicated and vaccination discontinued, our immunity to monkeypox virus has weakened.
(Magnification x125,000 at 10cm tall)

Right: Human parainfluenza virus, (transmission electron micrograph)

Parainfluenza is a common infection of babies and infants, affecting the ear, throat and chest and giving rise to pneumonia and croup. This artificially coloured image of a parainfluenza virus particle shows clearly its main elements. In the middle are pale blue strands of RNA, the virus's genetic material. Surrounding them is a white capsule of protein, which in turn is coated in blue-green protein spikes which identify and connect with targeted human cells to infect them.
(Magnification x20,500 at 10cm tall)

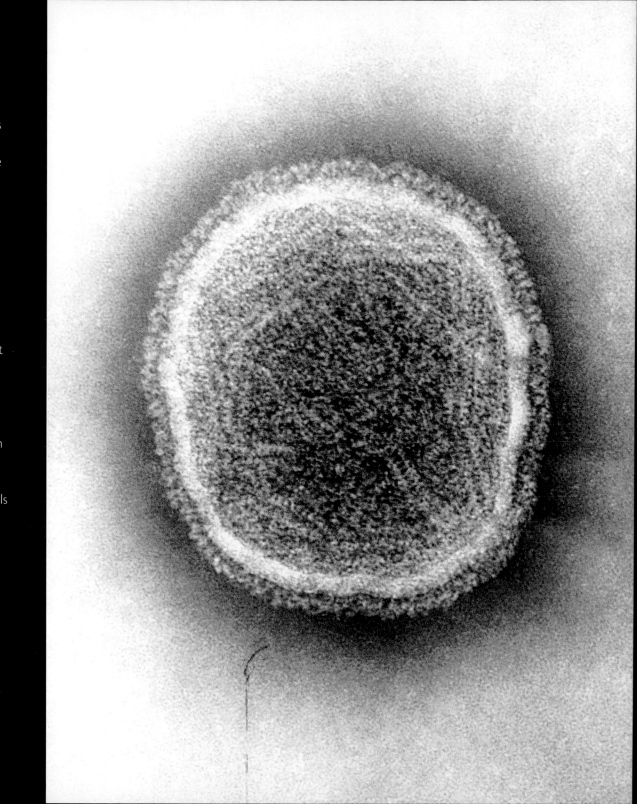

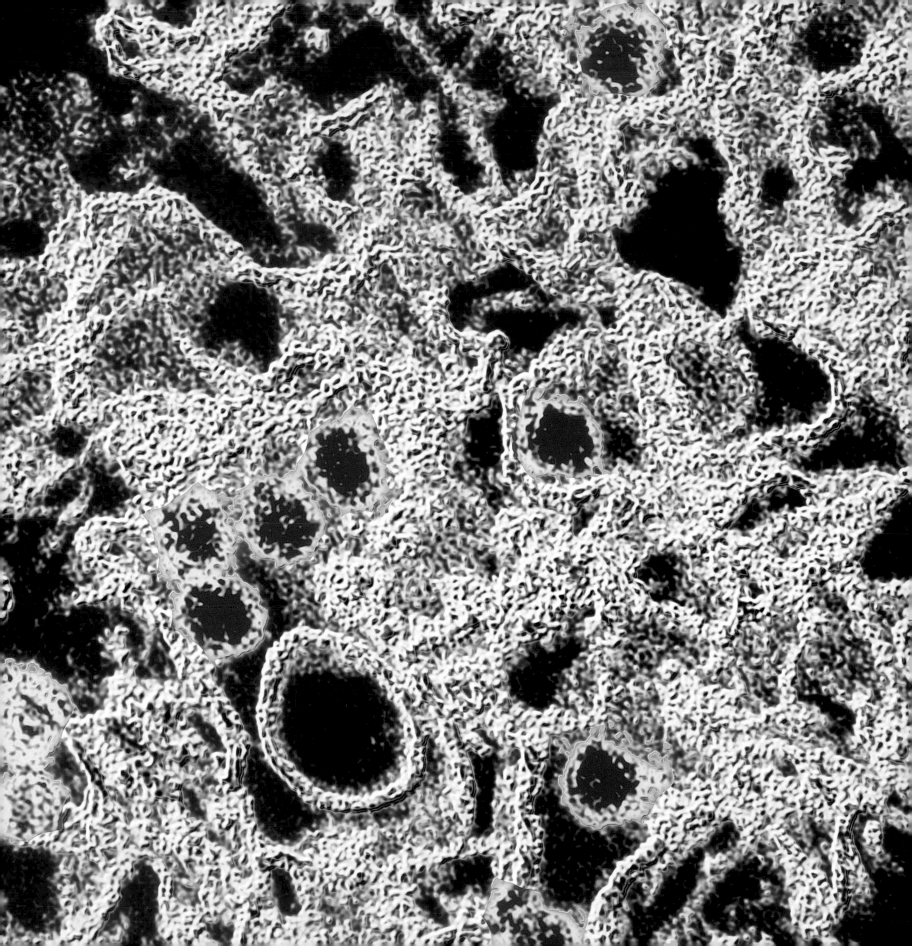

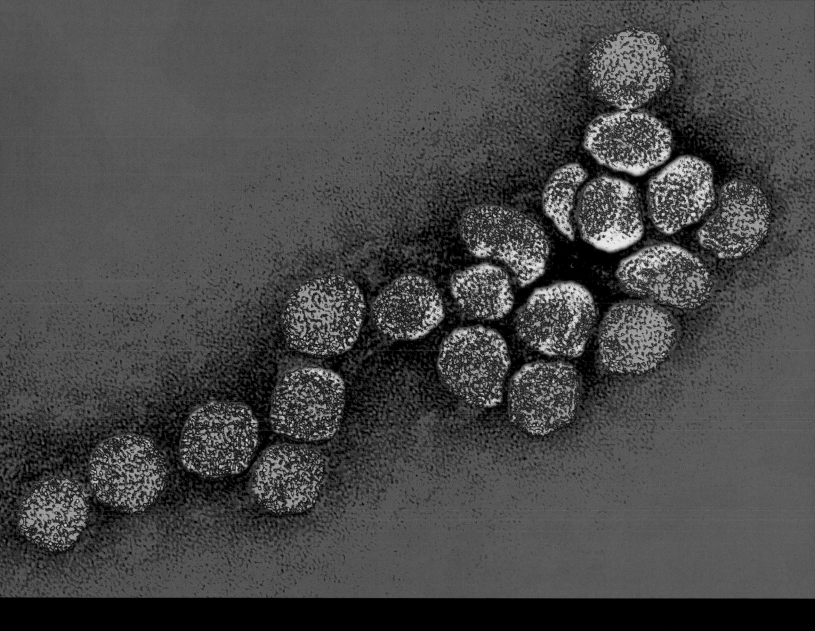

Left: Zika virus (transmission electron micrograph)
First identified in the 1950s, the Zika virus (named after a Ugandan forest) was found only in a narrow geographical band across equatorial Africa and Asia. Spread by mosquitoes, it crossed the Pacific Ocean to the Americas in 2007, where Zika became briefly epidemic in 2015. Typically its effects are relatively mild – fever, skin rash and joint pain – but pregnant women can infect their unborn children, resulting in serious birth defects. Trials of a vaccine began in 2016.
(Magnification unknown)

Above: The West Nile virus, a flavivirus (transmission electron micrograph)
Flaviviruses take their name from the Latin *flavus*, meaning yellow. The best known flavivirus causes yellow fever. Others in the family include the Zika virus, and most, including West Nile virus, are spread by mosquito bites. West Nile fever is not confined to the West Nile or even Africa. It typically includes muscle ache, skin rash, headaches and nausea. In extreme cases it can induce diseases such as encephalitis and meningitis, which cause inflammation of the brain or spinal cord.
(Magnification unknown)

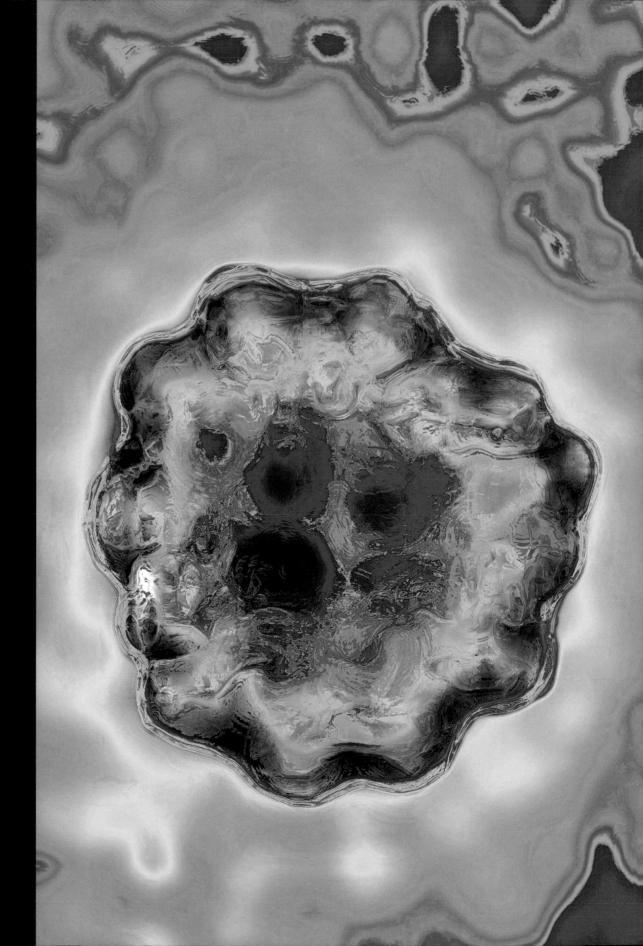

**Hepatitis D virus (HDV)
(transmission electron micrograph)**
Hepatitis (inflammation of the liver) may be the result of many things including excess of alcohol or other toxins. But the commonest cause is one of the five hepatitis viruses, identified as A, B, C, D or E. Hepatitis D is only contracted by patients already suffering from hepatitis B, a development called superinfection. Most patients recover completely, but if the infection becomes chronic it is almost certain to lead to liver failure and liver cancer. (Magnification x 2,850,000 at 10cm wide)

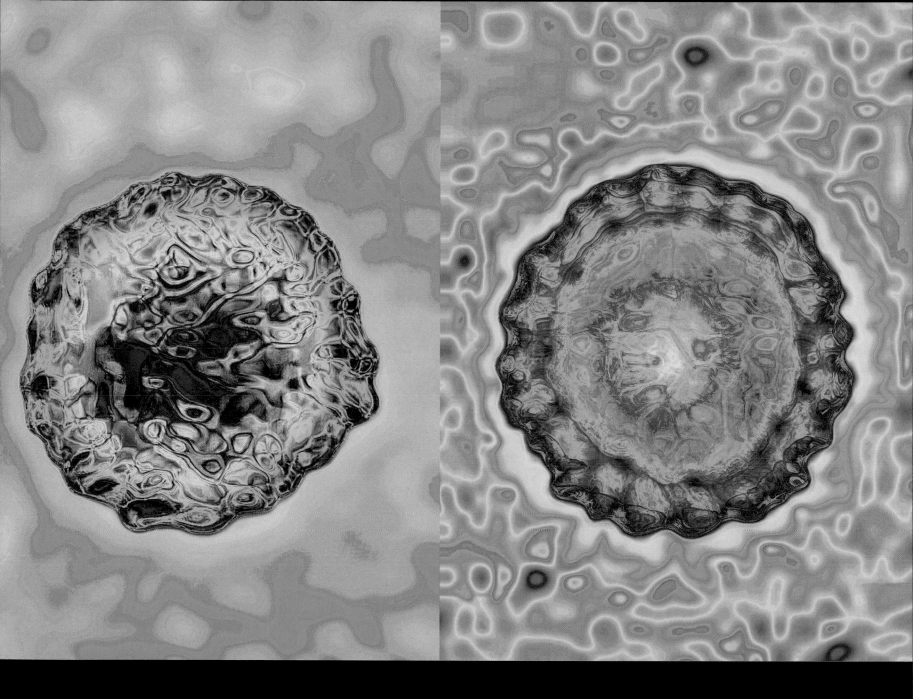

Above: Hepatitis C virus (HCV) (transmission electron micrograph)
Of the five hepatitis viruses, A and E are caught through infected food and water; the rest enter the liver via infected blood. Drug users are at great risk by sharing contaminated needles. Although vaccination is available for types A, B and D, there is none for C, which can become chronic. A lengthy course of oral medication cures most chronic cases. Untreated however, it is the commonest reason for liver transplants. (Magnification x1,800,000 at 10cm wide)

Above: Hepatitis B virus (HBV) (transmission electron micrograph)
Hepatitis B and C are both acquired through contact with contaminated blood, and B is most commonly transmitted sexually. Sometimes it is passed from mother to child in the womb, and infant sufferers are almost certain to develop a chronic hepatitis condition – infection after the age of five rarely does occur. In most adults, hepatitis B clears without medication; and no medication is completely effective in removing the virus, although drugs can reduce its effects.
(Magnification x4,500,000 at 10cm wide)

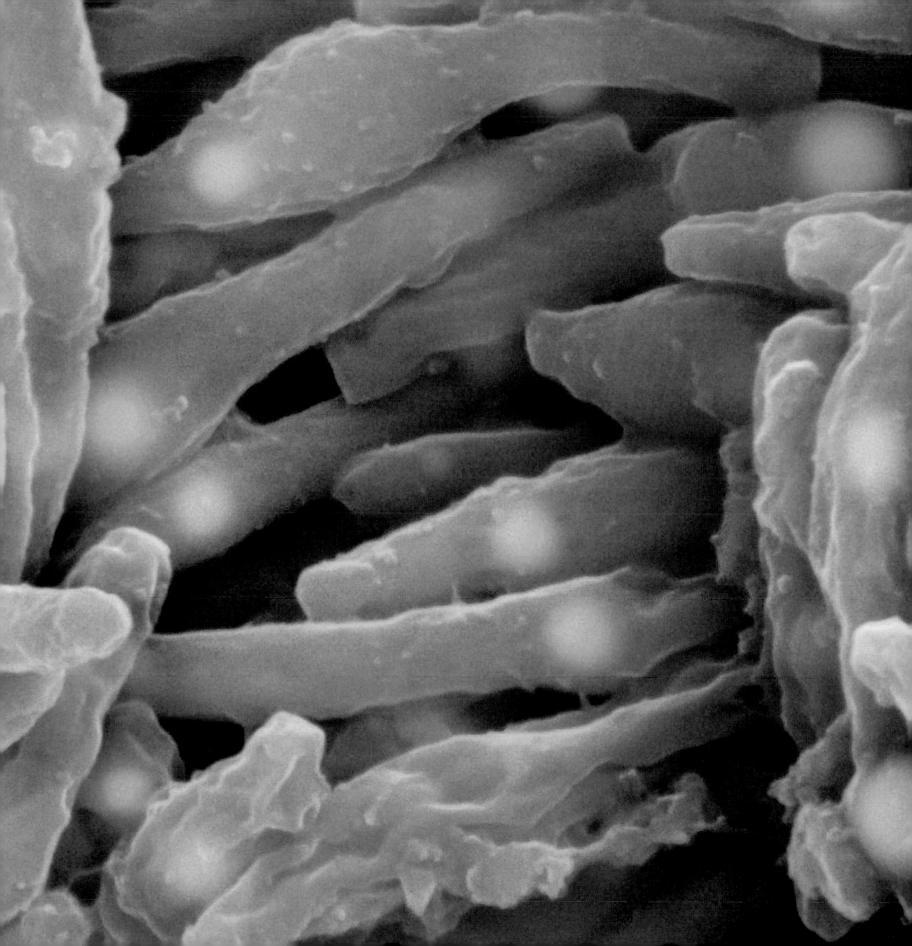

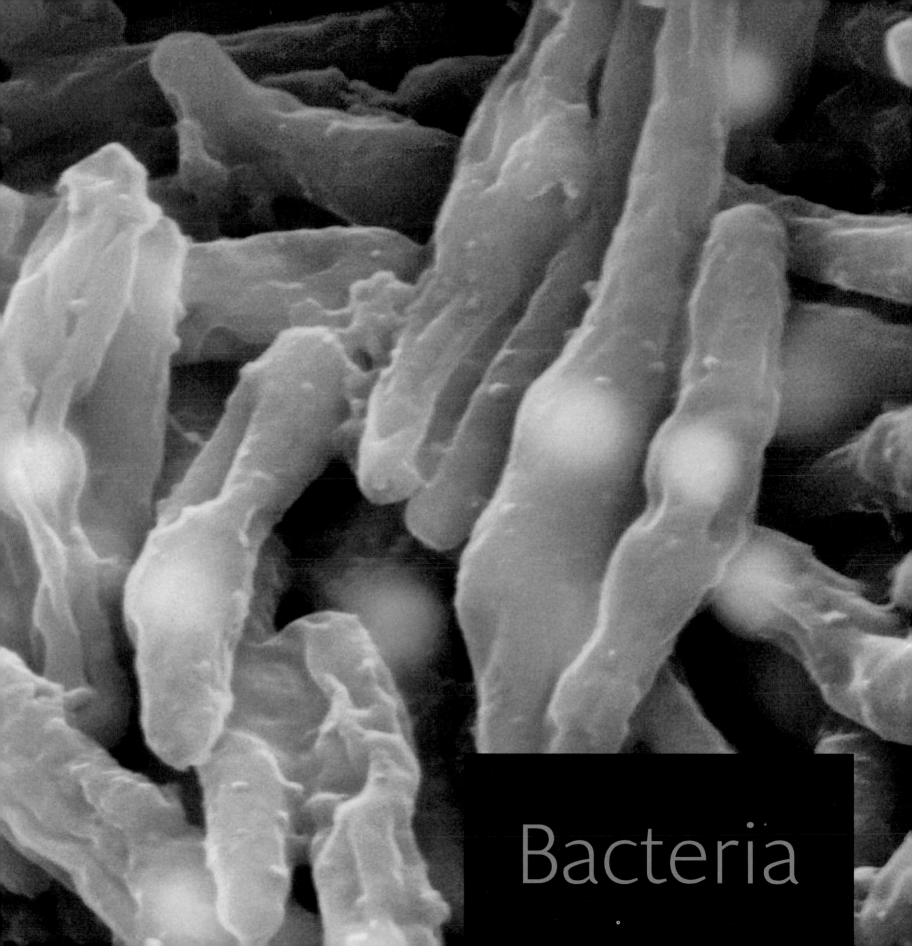

Bacteria

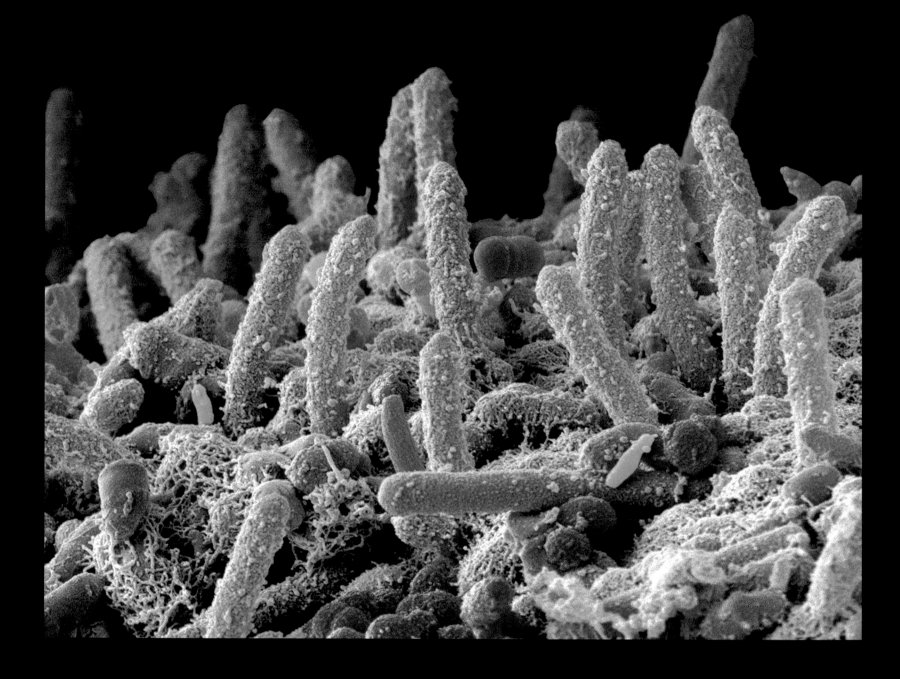

Previous page: Tuberculosis bacteria (scanning electron micrograph)
Bacteria come in many shapes and sizes. These elongated ones, for which the medical term is rods, give us tuberculosis. They are transmitted by coughs and sneezes, then inhaled. Although their main target is the lungs, they can also enter the bloodstream and infect other parts of the body. In the lungs they cause warts called tubercles, made of bacteria and dead tissue. Tuberculosis can be prevented by childhood vaccination and treated by antibiotics.
(Magnification x13,300 at 10cm wide)

Above and right: Oral bacteria (scanning electron micrograph)
These artificially coloured images show a variety of bacteria in the mouth – on the right, on the inner surface of the cheek and, on the left, in greater detail. A great many bacteria are to be found throughout the body, and indeed throughout the planet on land and sea, in the earth and in rivers. Many are beneficial to humans, and those that aren't can normally be defeated with antibiotics.
(Above: Magnification x6500 at 10cm tall)
(Right: Magnification x10,000 at 10cm wide)

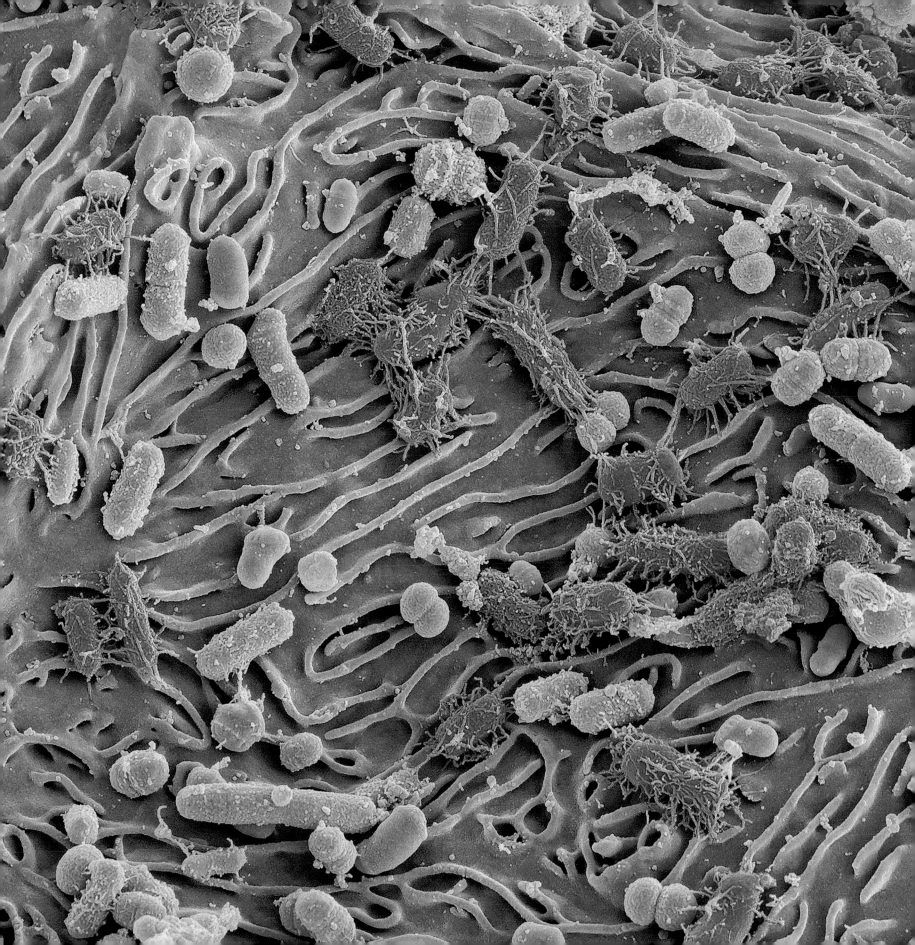

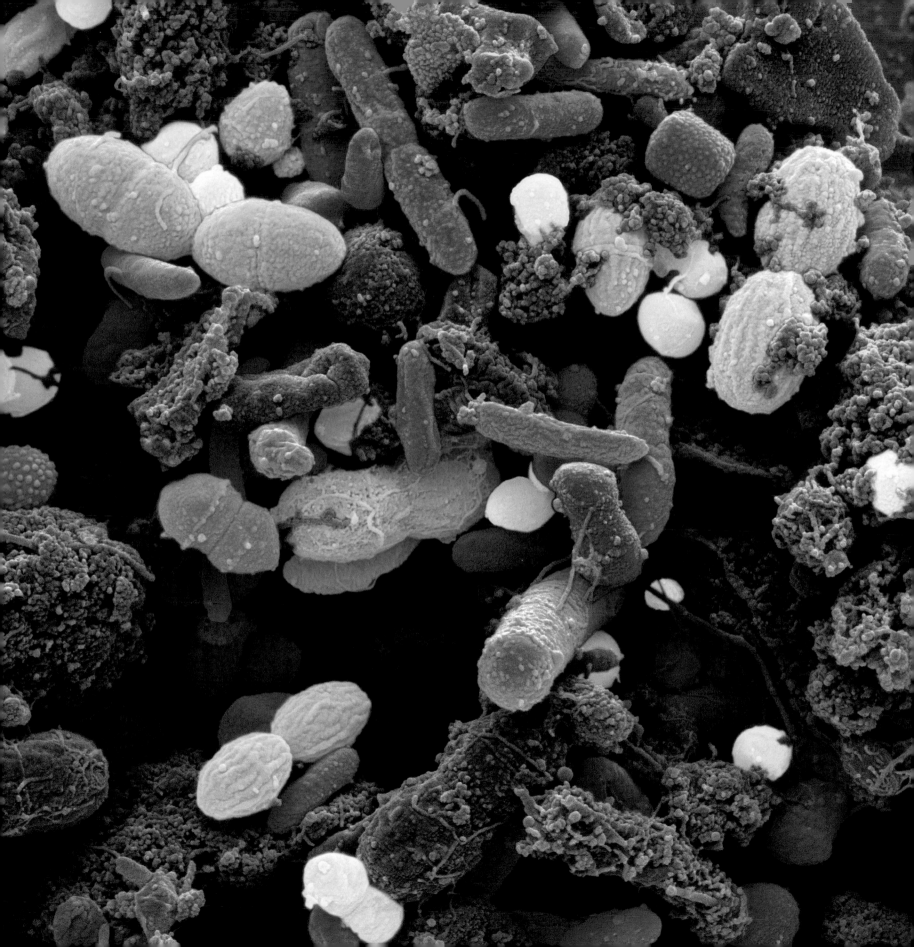

Left: Faecal bacteria (scanning electron micrograph)

At least half of all human faeces consists of bacteria. The human gut relies on many so-called 'good' bacteria which aid in the digestion of nutrition. Others such as salmonella and E. coli, which enter the gut in infected food, can cause serious illness. Because good and bad bacteria are expelled in faeces, good personal hygiene after defecation is an important way of restricting the spread of disease. (Magnification x8,000 at 10cm wide)

Right: Bacteria and yeast (scanning electron micrograph)

Bacteria and fungal often co-exist in nature, forming a mutually beneficial relationship called fungal-bacterial endosymbiosis. In this image the pink rods are bacteria and the shorter red ones are yeast, a form of fungus. Food producers exploit this symbiotic set-up, which is also of interest to medical researchers. For example, bacteria (E. coli and *Serratia marcescens*) and fungus (*Candida tropicalis*) are related factors in the development of Crohn's disease. Understanding their interaction may lead to better treatment. (Magnification x6,000 at 10cm tall)

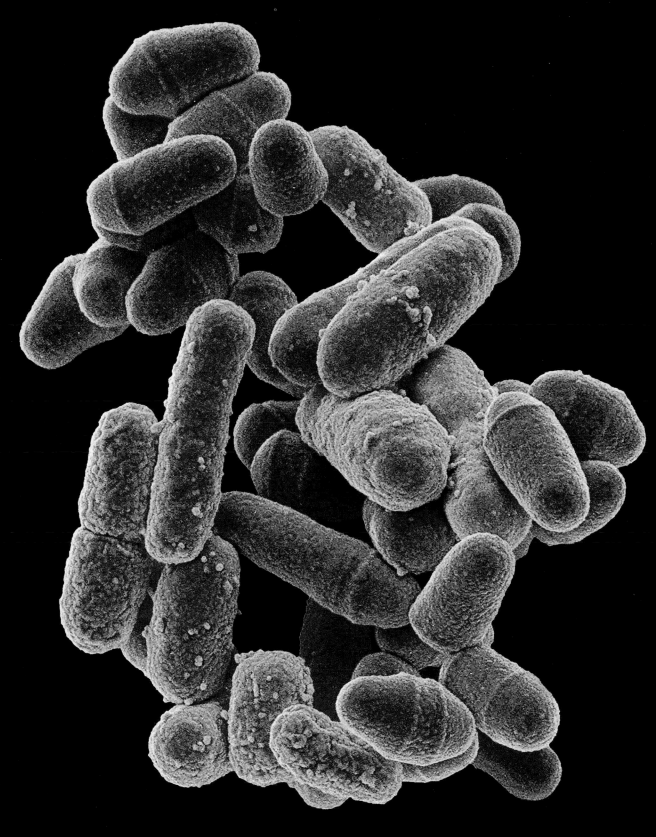

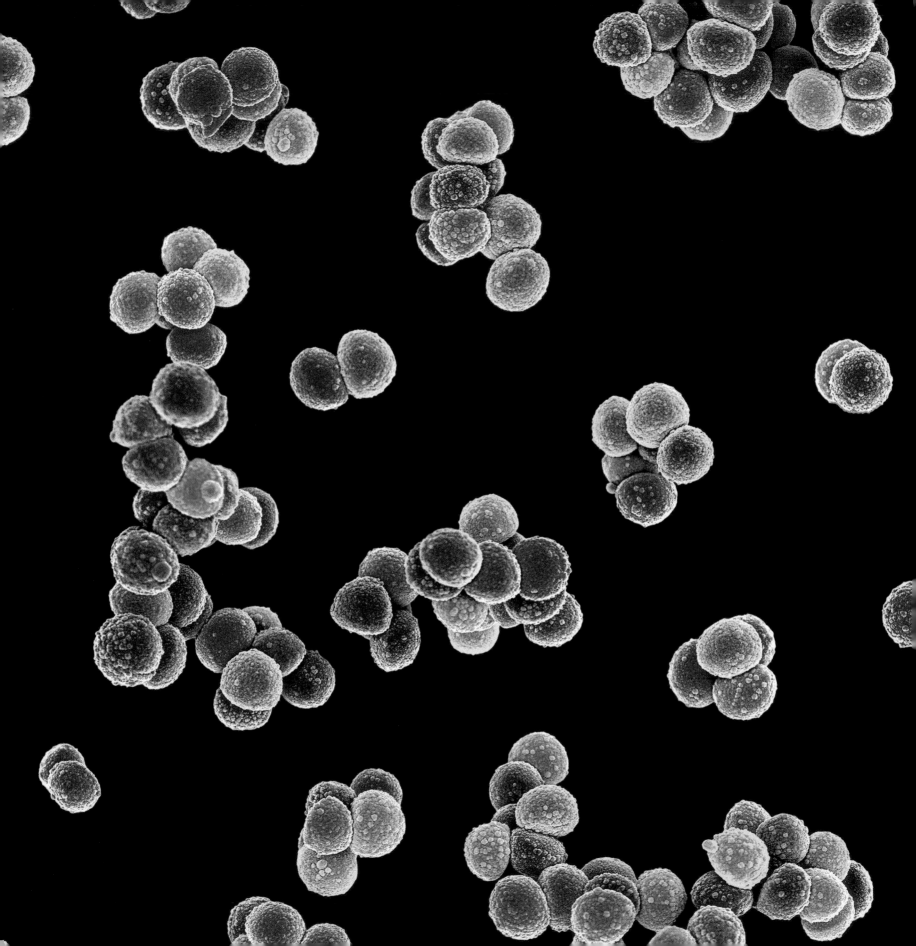

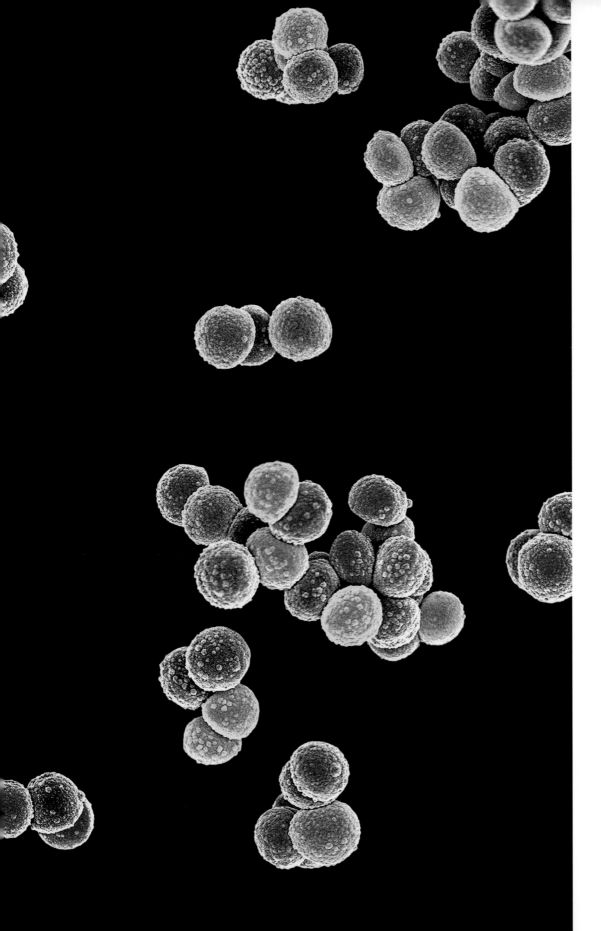

MRSA bacteria (scanning electron micrograph)

Methicillin-resistant *Staphylococcus aureus*, MRSA, has acquired a reputation as a hospital superbug. Although one in three of us carry these bacteria with no ill-effects thanks to a healthy immune system, not all are so fortunate. Those with weakened defences are vulnerable, particularly patients recovering from surgery or with invasive devices such as catheters. Boils on the skin are the first signs of infection, which can spread to other organs. MRSA is very resistant to antibiotics. (Magnification unknown)

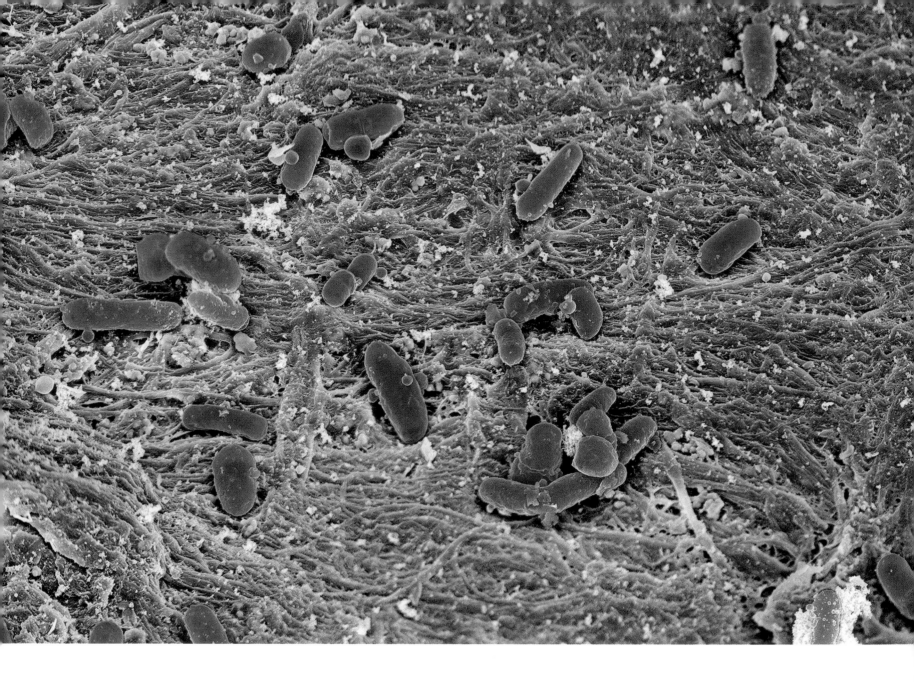

E. coli bacteria on a dollar bill

In the background of this false-colour image are the green fibres of a dollar bill, on which you can see purple rods of *Escherichia coli*, or E. coli bacteria. There are many strains of E.. coli, most of which live harmlessly in our guts, producing vitamin K2 and actually keeping out other, more harmful bacteria. A few however, entering our bodies on contaminated food, cause severe diarrhoea, fever and urinary infections. Treatment is by rehydration and antibiotics.

(Magnification unknown)

Helicobacter pylori in antral crypt

Two out of three of us carry *Helicobacter pylori* bacteria in our guts with
no ill effects at all. But by attacking the acid-resistant mucus that lines
the stomach wall they are responsible for most stomach ulcers. Although
ulcers tend to appear in adults, the bacteria are most commonly acquired
in childhood through dirty water. Here you can make out the bacteria as
small dark specks in the space within the folds of the stomach lining.
(Magnification unknown)

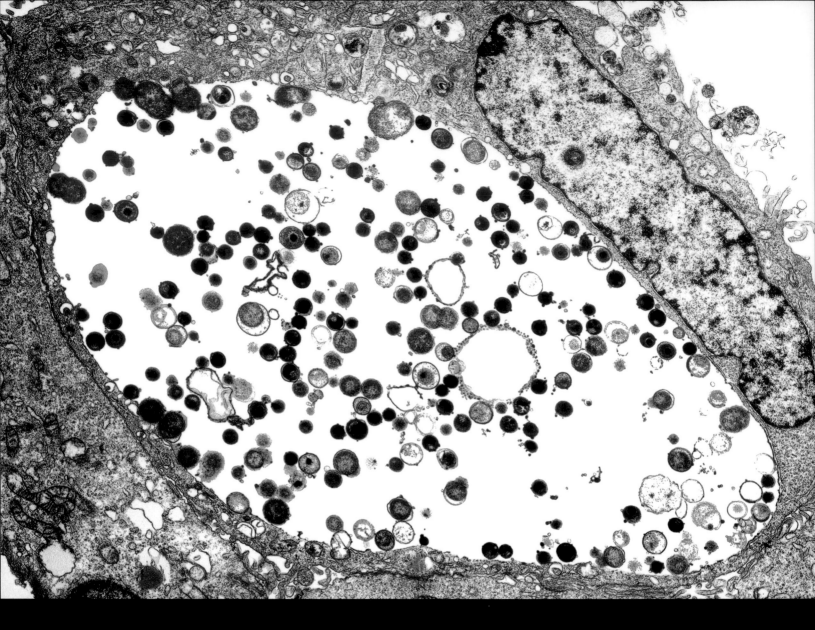

Above: *Chlamydia trachomatis* bacteria (transmission electron micrograph)
The blue spheres here are chlamydia bacteria within an infected cell (red). The bacteria rely on the amino acids of cells for nutrition. Chlamydia is a sexually transmitted bacterium which causes diseases of the genitals, eyes and lymph nodes. In untreated cases it can lead to blindness, but is easily treated with antibiotics. It is three times more common in women than in men, and regular screening is recommended for women under 25.

Right: Gonorrhoea bacteria (scanning electron micrograph)
This image shows gonorrhoea bacteria (red) on a human skin cell (green). Gonorrhoea is a sexually transmitted disease with symptoms similar to chlamydia – genital pain and discharge. The condition's common name, the clap, is said to come from the French slang word for a brothel where such a disease might have been caught – *le clapier*. Left unchecked, gonorrhoea can spread through the body and inflame joints and heart valves.

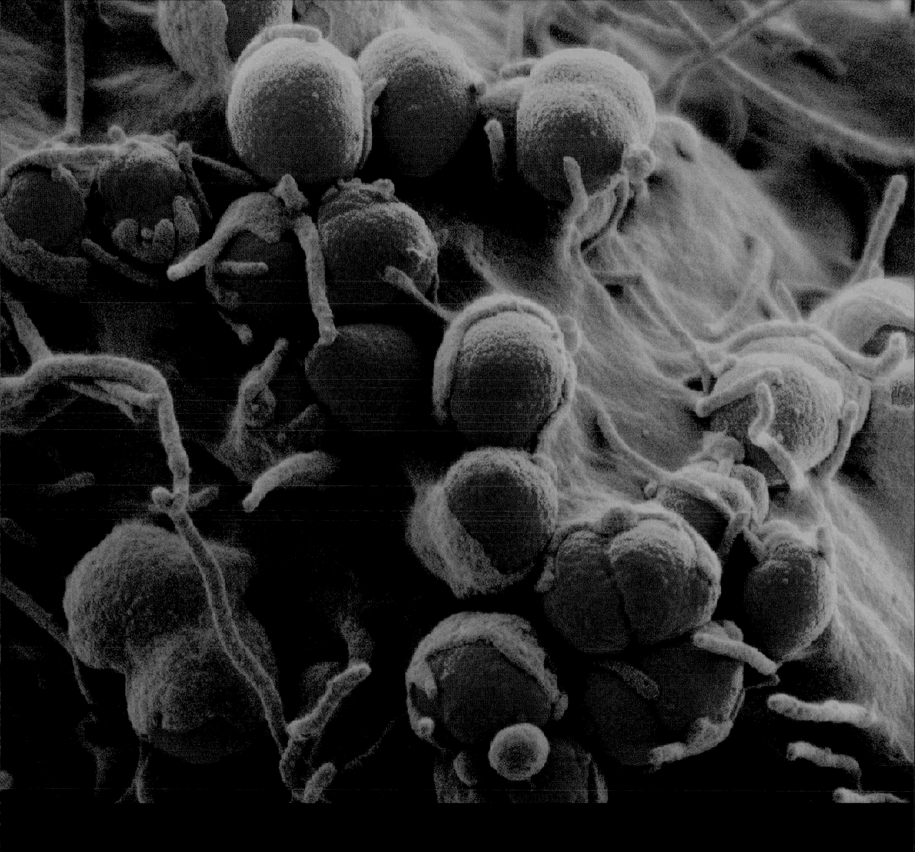

Bacteria 91

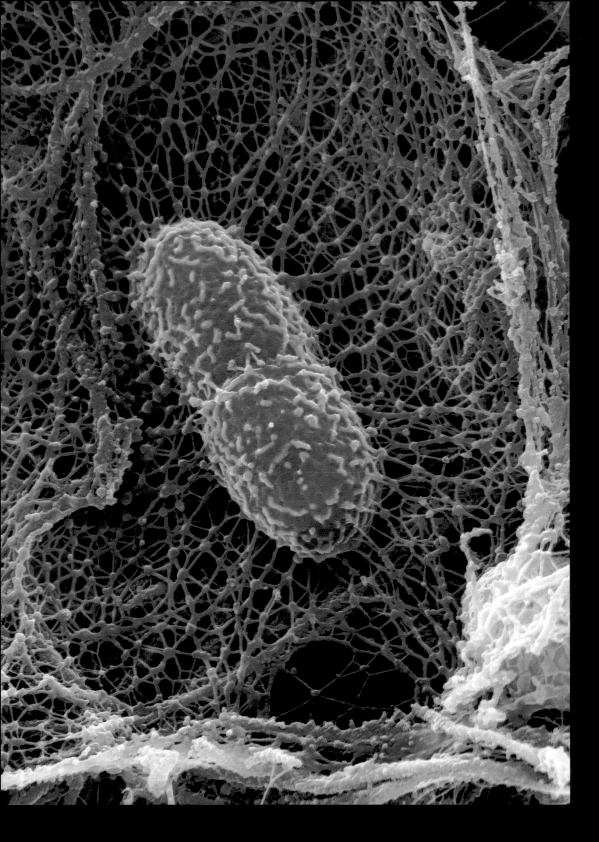

Left: *Klebsiella pneumoniae* bacterium (scanning electron micrograph)

Klebsiella pneumoniae bacteria can cause pneumonia in patients with a weakened immune system. This picture however shows a body's defences working well against a bacterium. In this case the body is that of a mouse, and the bacterium (pink) is attacking the mouse's lung tissue. Neutrophils, the most plentiful cells in the immune system, respond by spinning a complex web (green) of DNA, RNA and protein, which traps and kills the invader. (Magnification unknown)

Right: *Staphylococcus epidermidis* bacteria (scanning electron micrograph)

This is one of 40 species of Staphylococcus bacteria. The name literally means 'bunch of grapes' (*Staphylococcus*) 'on the skin' (epidermidis). Like its sister bacterium MRSA, *S. epidermidis* is common on human skin and is usually harmless. However, people with low immunity, open wounds or medical implants such as catheters, are vulnerable to *S. epidermidis*. It has the remarkable ability to create a biofilm that binds to plastic, and to which other bacteria can then bind.
(Magnification x5,800 at 6cm wide)

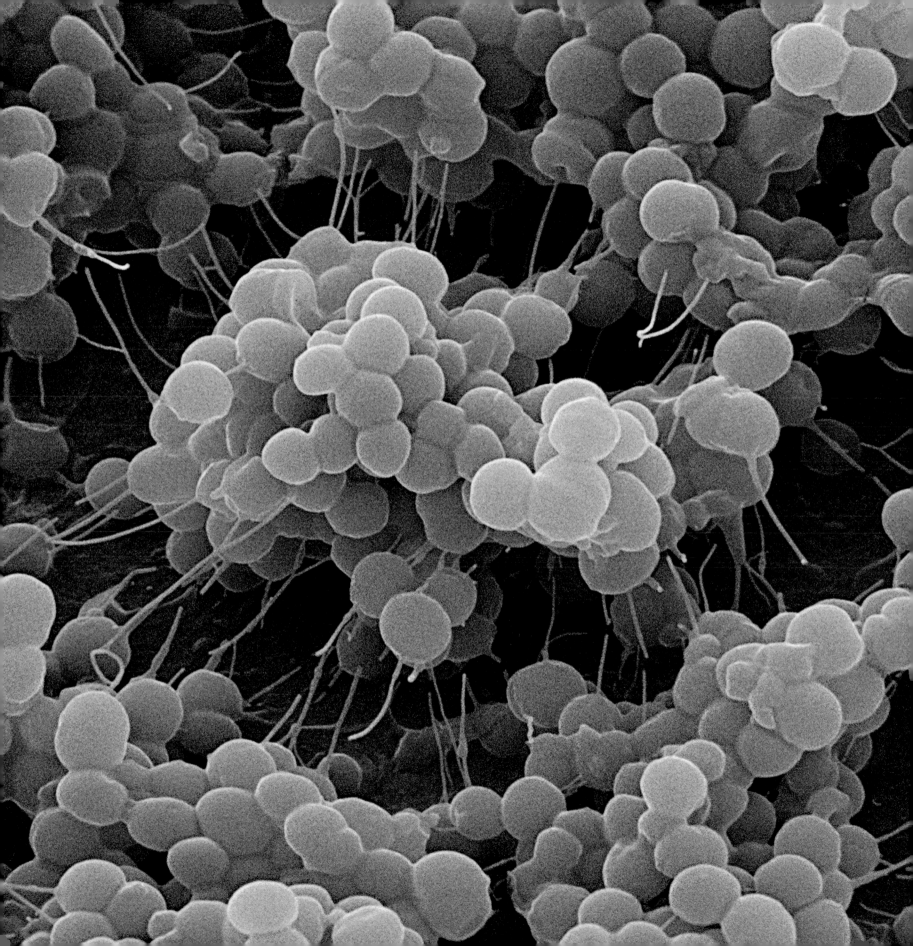

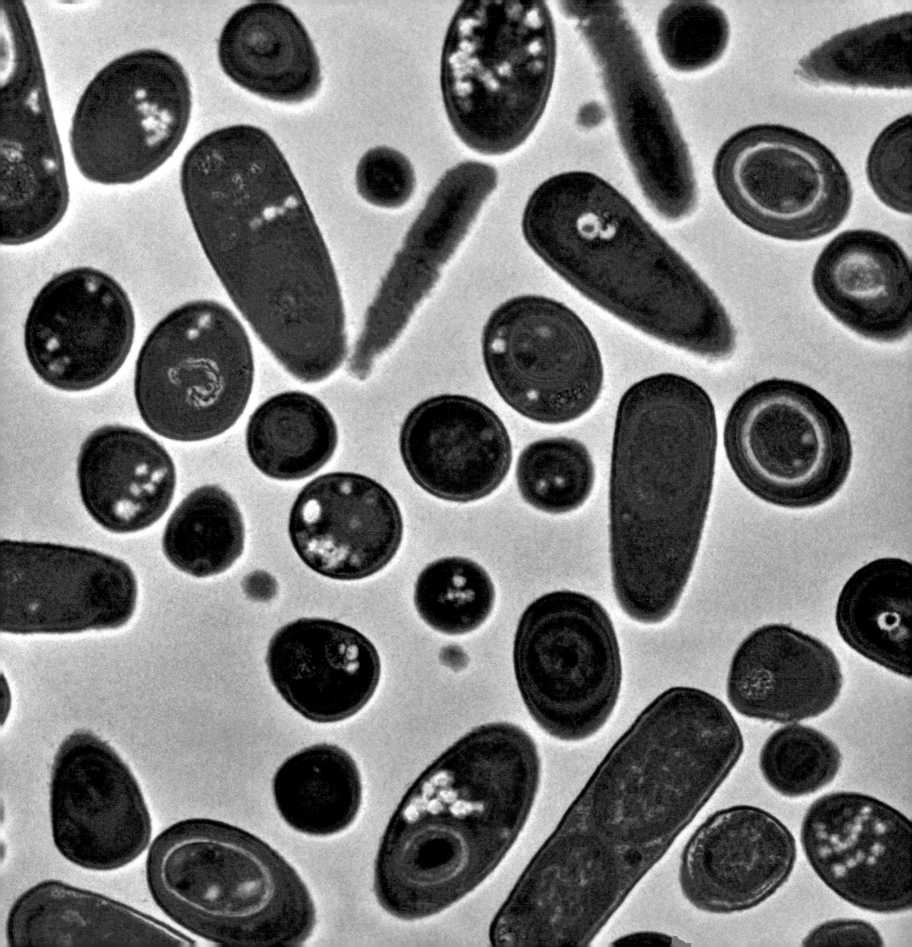

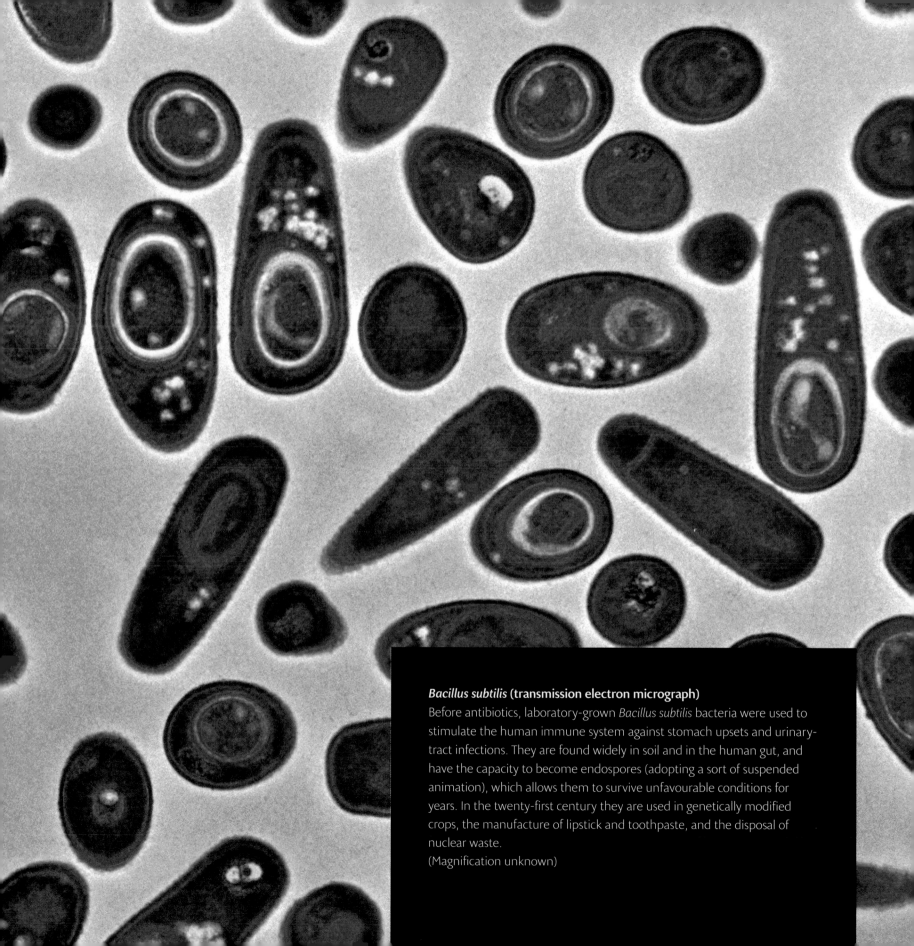

Bacillus subtilis **(transmission electron micrograph)**
Before antibiotics, laboratory-grown *Bacillus subtilis* bacteria were used to stimulate the human immune system against stomach upsets and urinary-tract infections. They are found widely in soil and in the human gut, and have the capacity to become endospores (adopting a sort of suspended animation), which allows them to survive unfavourable conditions for years. In the twenty-first century they are used in genetically modified crops, the manufacture of lipstick and toothpaste, and the disposal of nuclear waste.
(Magnification unknown)

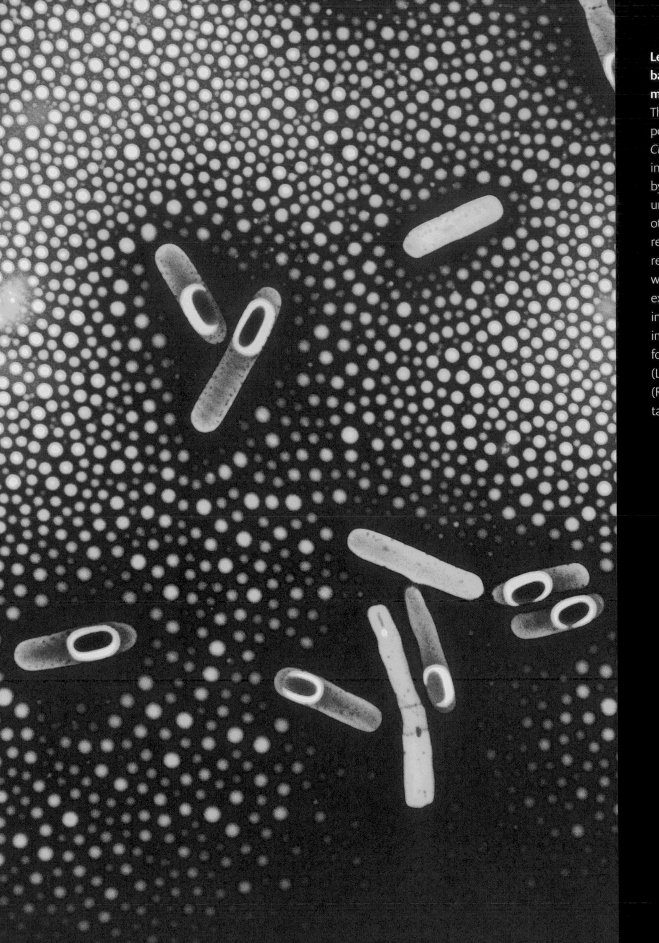

Left and right: *Clostridium difficile* **bacteria (transmission electron micrograph)**
The healthy gut contains a balanced population of many bacteria, including *Clostridium difficile* (seen in these two images). That balance can be disrupted by antibiotics (prescribed for some unrelated infection), which kill off other bacteria, allowing antibiotic-resistant *C. difficile* to dominate. The resulting inflammation of the colon, with severe diarrhoea, can be fatal in extreme cases. The dark purple spots inside the bacteria in the left-hand image are spores, which can survive for weeks on uncleaned surfaces.
(Left: Magnification x2,500 at 10cm tall)
(Right: Magnification x32,000 at 10cm tall)

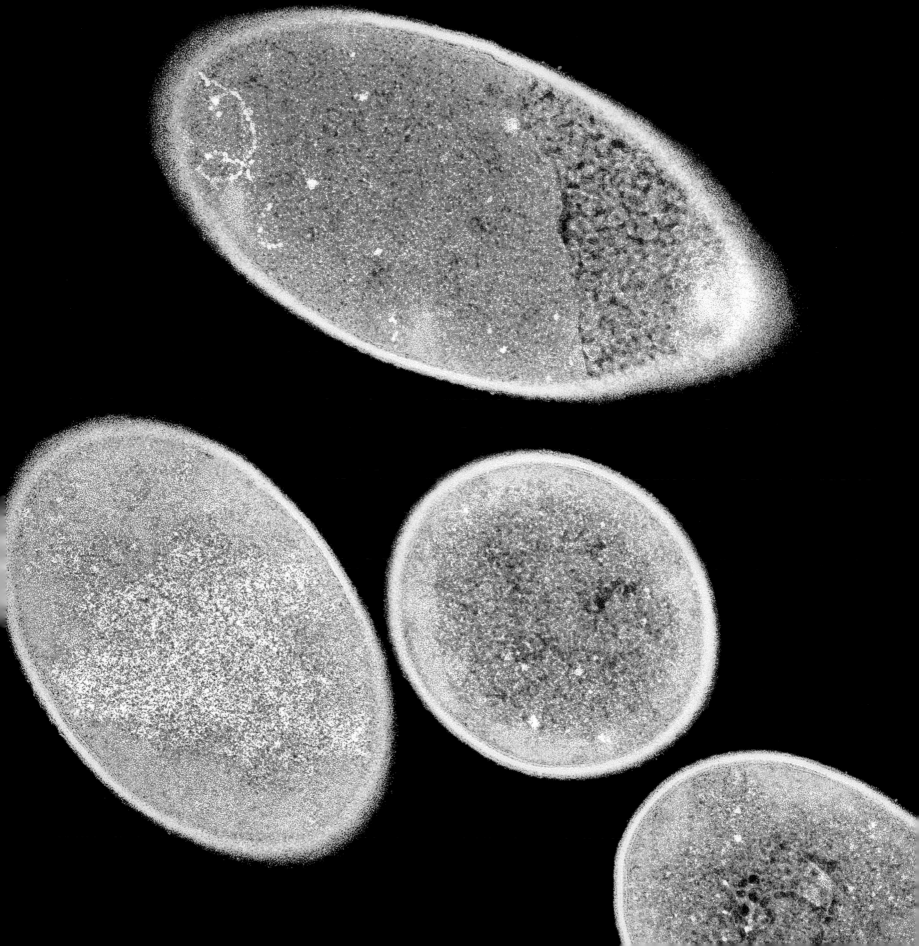

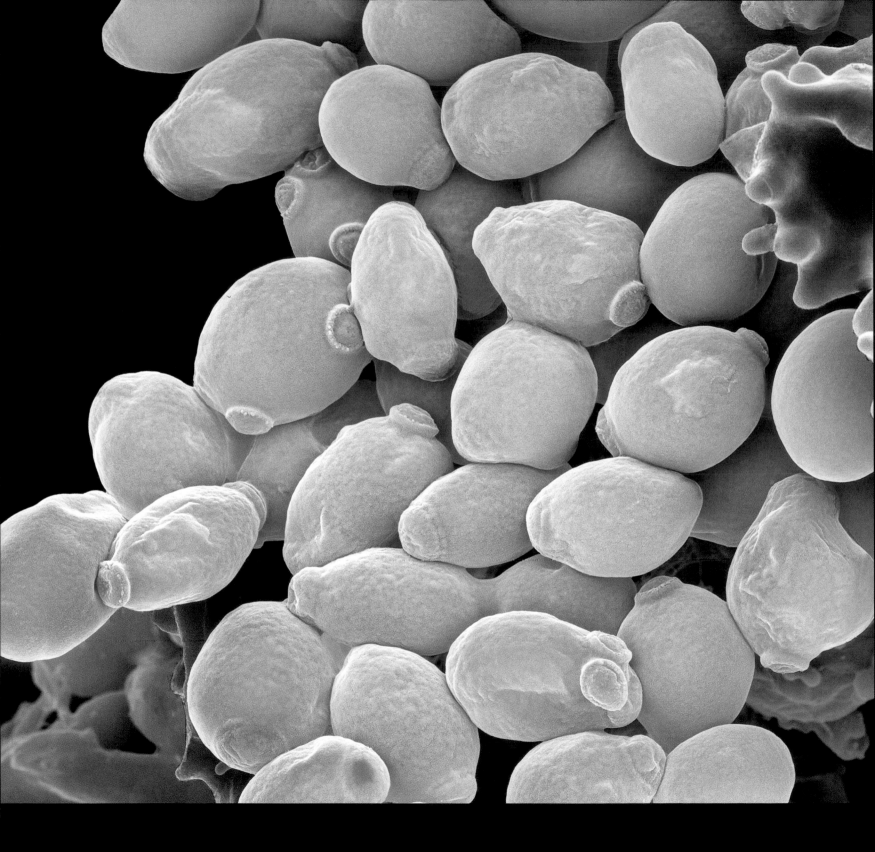

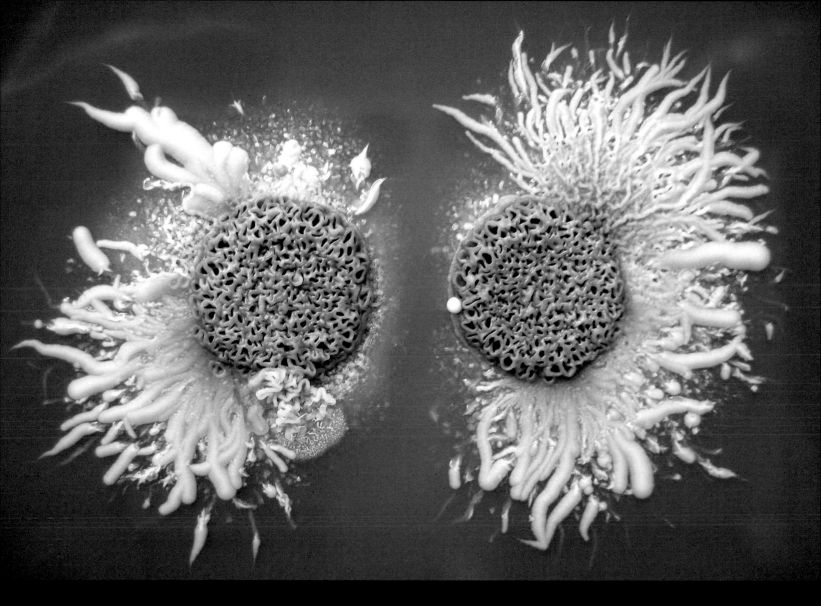

Left: *Candida albicans* yeast cells (scanning electron micrograph)
Candida albicans is a yeast fungus commonly found in many humans with no ill effects. Sometimes a course of antibiotics or a weakened immune system can allow its growth to accelerate, resulting in oral or vaginal thrush, or nappy rash. This excessive growth triggers a change in *Candida albicans*: the normally single-celled fungus becomes multicellular and resistant to antifungal and antibiotic treatments. This image shows *Candida albicans* present in the urine sample of someone with a urinary tract infection.
(Magnification x4,000 at 10cm wide)

Above: *Candida albicans* fungus (macro photography)
This image shows two *Candida albicans* colonies grown in laboratory conditions. They are growing on agar, a material derived from algae, which is normally a dull green colour. *Candida albicans* has the effect of raising the pH levels of its surroundings – in other words, making them less acid and more alkaline. In the process the agar also turns blue. The white extremities of the colonies reach out into the agar for nutrition, but not towards their neighbouring colonies.
(Magnification unknown)

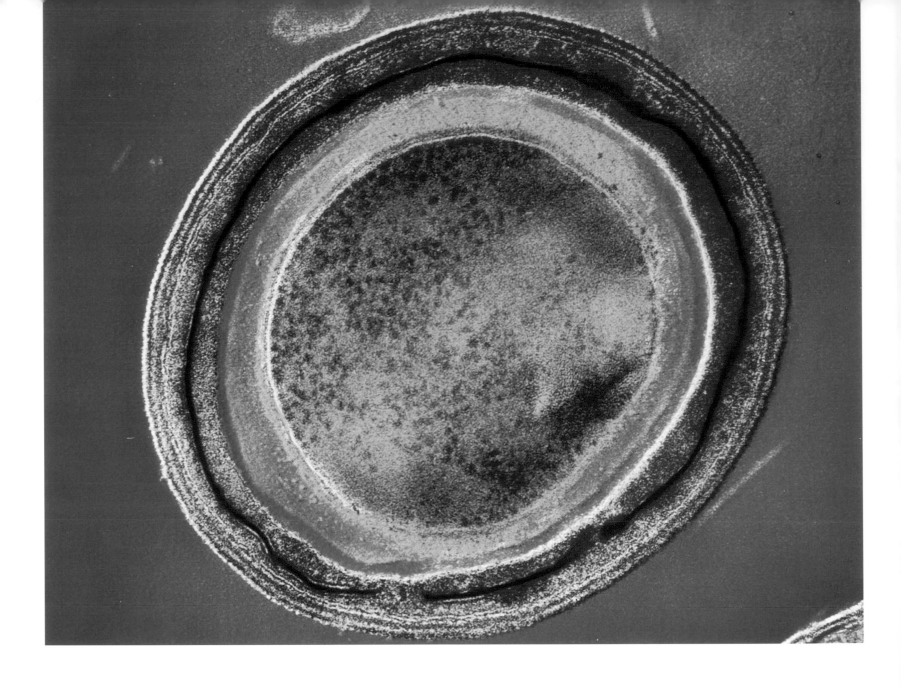

Above: *Clostridium tetani* **bacterial spore (transmission electron micrograph)**

Many bacteria can enter a spore state – a sort of protected hibernation until conditions become more favourable for germination. In this image a spore of *Clostridium tetani* bacterium (orange and green) is encased in a coating of many membranes (purple). *C. tetani* exists in soil and in the intestines. It is harmless until it enters a wound, when it causes tetanus, also known as lockjaw – a painful, sometimes fatal condition of muscle spasms. Vaccination against tetanus is available.
(Magnification x12,150 at 10cm wide)

Right: *Bordetella pertussis* **bacteria (transmission electron micrograph)**

This micrograph shows, in cross-section, the bacteria that give us whooping cough. The yellow matter in the middle of each one is the bacterium's DNA, surrounded by a protective cell wall. Whooping cough is named after the sound made by inhaling at the end of a coughing fit. The bacteria infect the air passage and are potentially fatal to new babies, especially premature ones. There is an effective preventative vaccine, but antibiotics have limited success after infection.
(Magnification x9,000 at 6cm wide)

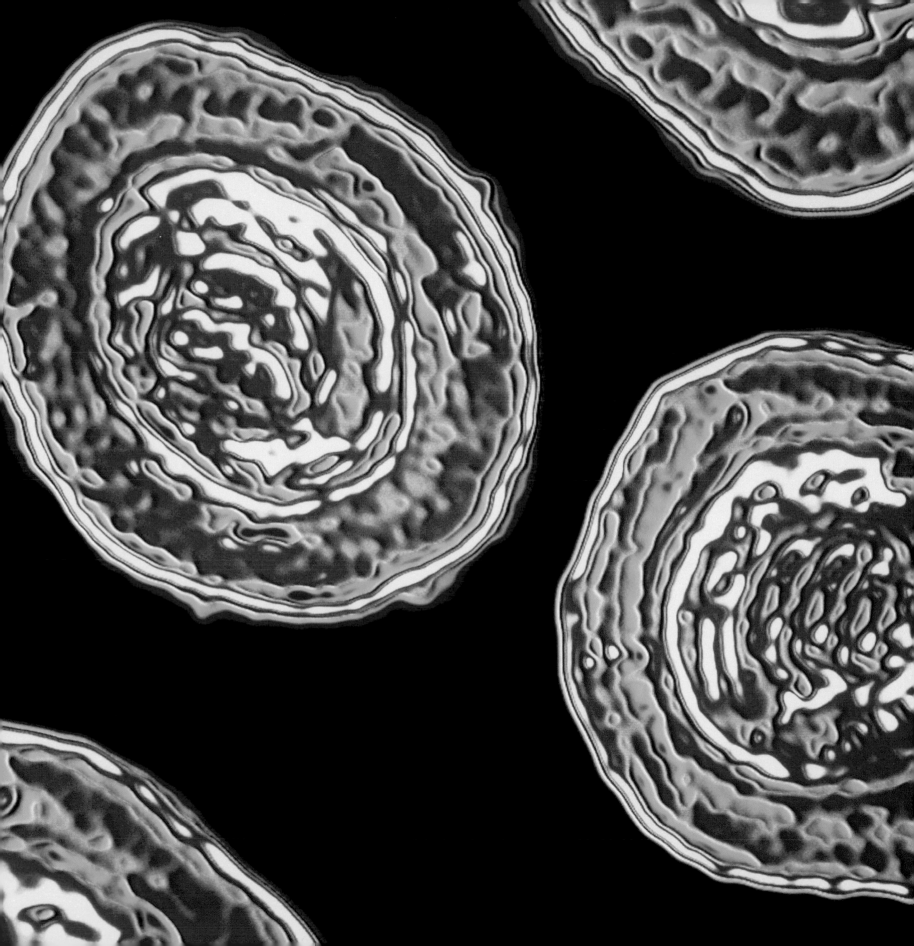

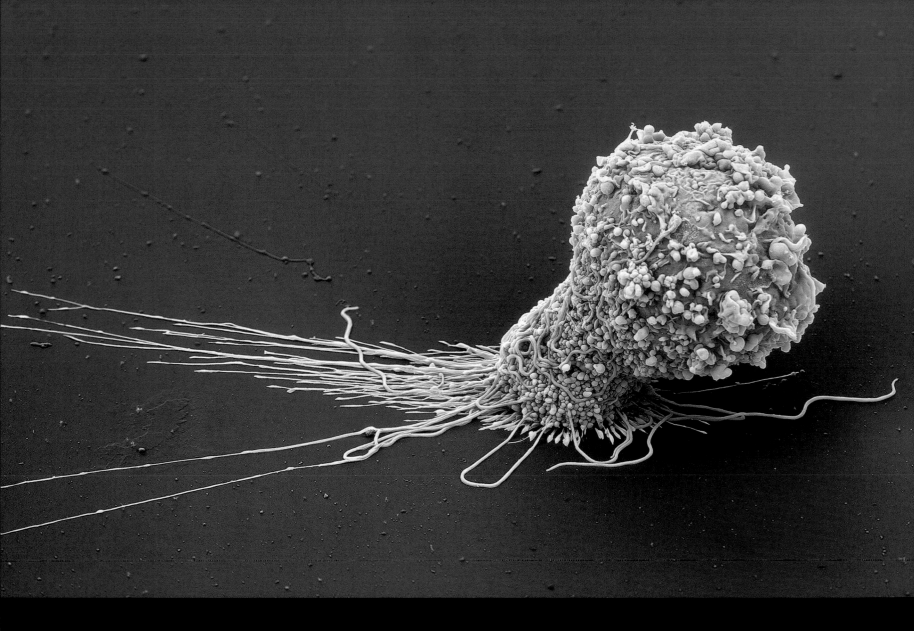

Above: Macrophage engulfing bacteria (scanning electron micrograph)
Macrophages are white blood cells that provide our defence against bacteria. Their job is to find and destroy potentially harmful bacteria in the bloodstream. In this image a macrophage (here yellow) has used its long tendrils to trap and reel in a *Borrelia* bacterium (just visible, here blue). *Borrelia burgdoferi* are caught from ticks or lice and can give us Lyme disease. Having captured it, the macrophage then engulfs its enemy, disabling any threat it presented.
(Magnification x2080 at 10cm wide)

Right: MRSA ingestion by white blood cell (scanning electron micrograph)
The rough yellow spheres in this image are methicillin-resistant *Staphylococcus aureus* (MRSA) bacteria. They take advantage of weakened immune systems such as those of patients recovering from surgery in hospital or suffering from AIDS. MRSA is very resistant to antibiotics but no match for a healthy immune system. Here, in a process called phagocytosis, you can see MRSA bacteria being captured and swallowed up by a white blood cell. Such cells are the front line of the body's immune system.
(Magnification unknown)

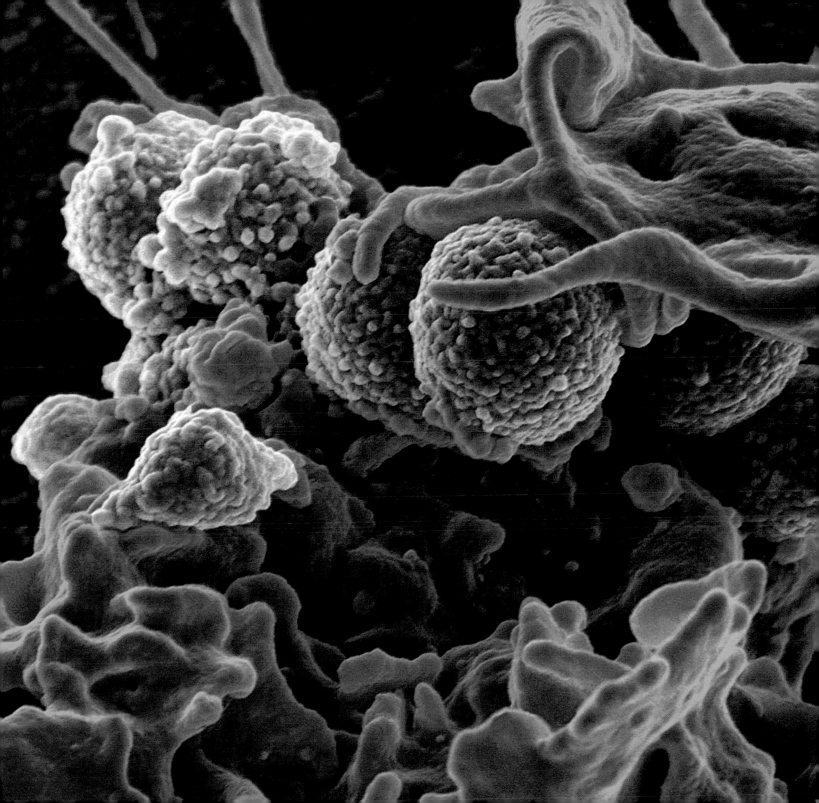

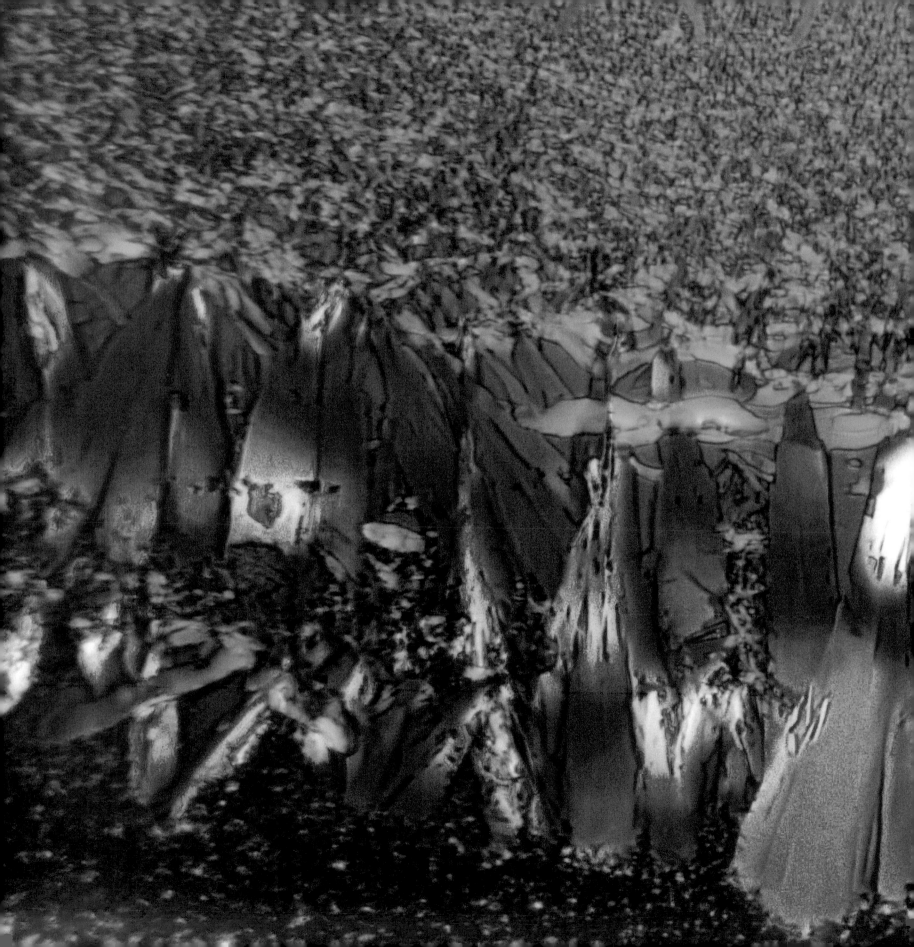

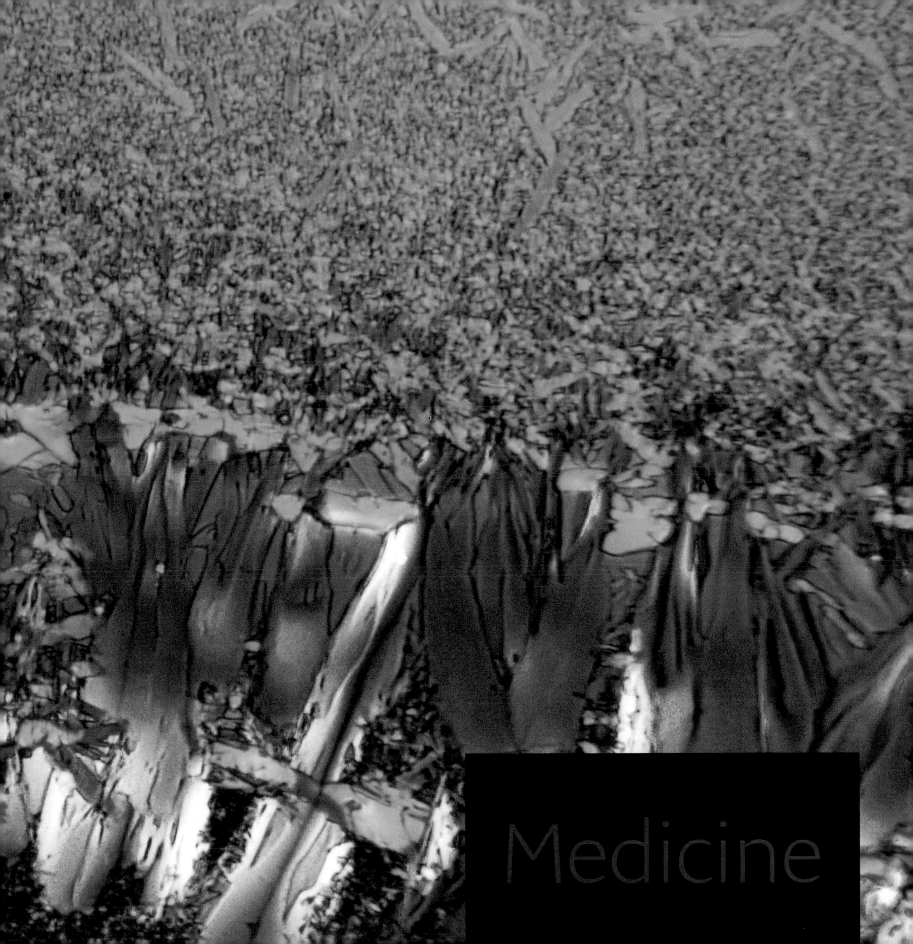

Medicine

Previous page: Penicillin (polarized light micrograph)

Penicillin was one of the earliest antibiotics to be harnessed in the fight against infectious staphylococcal and streptococcal bacteria, although it was some fourteen years after its discovery (by Alexander Fleming in 1928) that is was first used medically. Originally derived from the penicillium fungus, it is now a large synthetically produced family of medicines. But as its applications have grown so has the resistance of

Above: Beclometasone crystals (scanning electron micrograph)

Beclometasone is a corticosteroid, a synthetic version of a substance that is naturally produced by the adrenal cortex in response to stresses on the body. It is the active ingredient in the oral sprays that asthma sufferers use to relieve inflammation in their lungs. In a nasal spray it also acts to reduce enlarged nasal polyps, which can trap and prolong infections in the nose.

Embryonic stem cells (scanning electron micrograph)
Embryonic stem cells are the building blocks of our bodies. Created at
the very start of human life, they have a quality called pluripotency: from
identical beginnings they can develop into any of the different cells of
which the body is composed, depending on which biochemical signals
they receive in the womb. In theory it may be possible to engineer them to
repair tissue damaged by illness, but stem-cell research destroys embryos
and is therefore controversial.
(Magnification x1500 at 10cm wide)

Cocaine (polarized light micrograph)
Cocaine, derived from the leaves of the coca plant, still has some genuine medical uses, chiefly as a numbing agent for operations in the nose or mouth. Its effect in narrowing blood vessels also reduces bleeding. Its use as a so-called recreational drug is because it reduces the re-uptake of serotonin, dopamine and norepinephrine. The higher concentrations of these neurotransmitters left in the brain as a result gives users a brief sense of greater awareness and happiness.
(Magnification unknown)

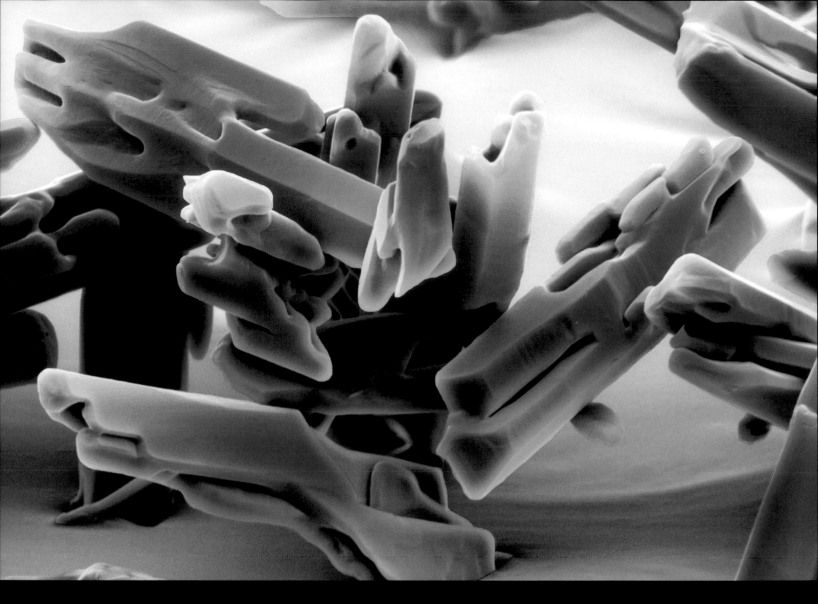

Above: Morphine crystals (scanning electron micrograph)
Morphine's misuse for recreational purposes creates a sense of detached euphoria; but as with cocaine the body develops addiction and tolerance so that higher and higher doses are required. Medically, morphine is still invaluable, working directly on receptors in the nervous system that are normally triggered by endorphin (the body's own natural painkiller). It is regularly used to relieve pain after a heart attack or during the labour of childbirth.
(Magnification unknown)

Right: Oxytocin crystals (polarized light micrograph)
Oxytocin occurs naturally in women, produced in the hypothalamus and secreted by the pituary gland. It has a remarkable range of social effects, influencing romantic attraction and sexual behaviour, and enhancing bonding within groups as well as between mother and child. Physiologically it is greatly involved in childbirth, sending signals to the foetal brain, causing contractions and even triggering lactation. Synthetic oxytocin is used medically to induce birth, release the placenta and encourage milk production.
(Magnification unknown)

Above and right: Insulin crystals (polarized light micrograph)
Insulin is a protein produced in the pancreas and its job is to control the levels of blood sugars. A deficiency of insulin results in a build-up of glucose in the blood. The result is diabetes, which is treated with injections of medical insulin. It is usually derived from the pancreas of pigs or cows, and is also used to treat high levels of potassium in the blood, which can cause palpitations and weakness of muscles.
(Magnification unknown)

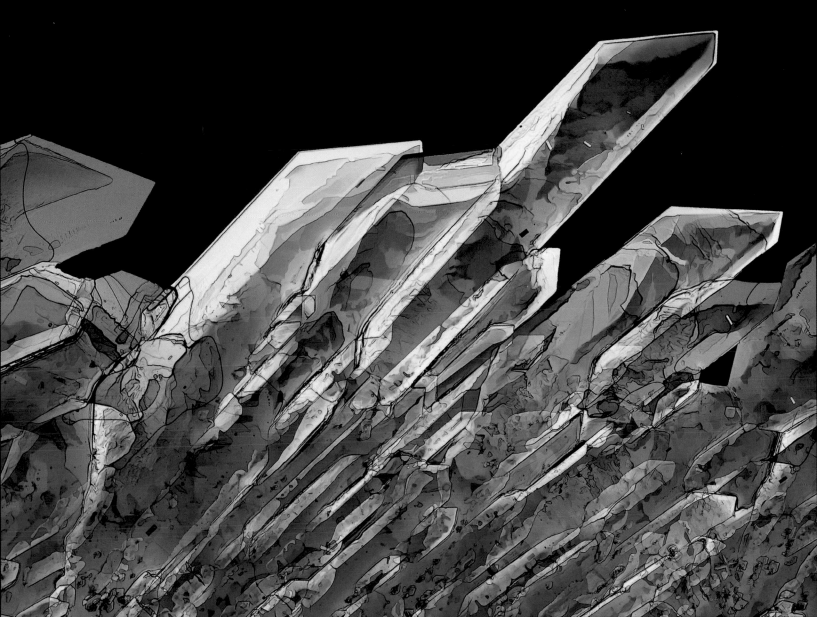

Cortisol (polarized light micrograph)
The word steroid simply means an organic compound with a particular molecular structure. Cortisol is a steroid hormone, produced by the adrenal gland, which works with adrenaline to rebalance the body when it is faced with stress. In the longer term it also acts to repair tissue, reduce inflammation and improve the body's defences against infection. The medical version, called hydrocortisone, is administered to patients with inflamed injuries or rheumatic conditions.
(Magnification x24 at 7cm wide)

Above and right: Testosterone hormone (polarized light micrograph)

Testosterone is the hormone that creates male characteristics in the
human body. When it interacts with an enzyme called 5-alpha-reductase
it becomes 5-alpha-dihydrotestosterone (known as DHT, and pictured
right), a more powerful version of the hormone. Testosterone occurs
in both males and females, but twenty times more is produced and
consumed by men. It is responsible for male genitalia and other attributes
such as muscle tone and hairiness. Medicinal testosterone is prescribed to
counter low natural production.
(Above: Magnification x100 at 7cm wide)
(Right: Magnification x16 at 7cm size)

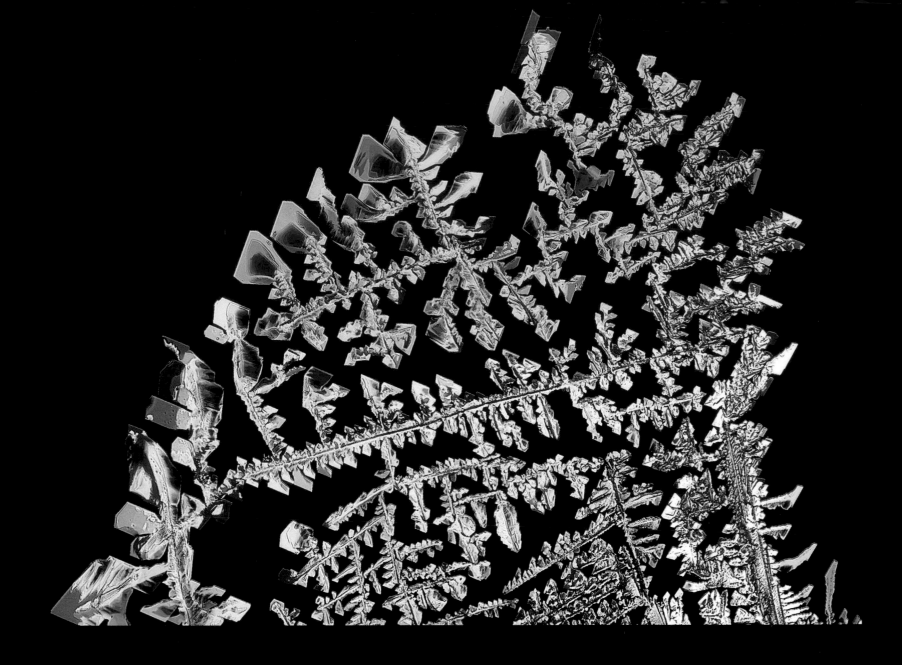

Above and right: female sex hormone (polarized light micrograph)
Oestradiol is the hormone which creates female characteristics in the
human body. It's the most potent of the six oestrogens and controls the
development of female genitalia and other attributes. In medicine it acts
as hormone replacement therapy for menopausal women, and for male-
to-female gender reassignment. It has been used to treat both breast and
prostate cancer. Oestradiol also has applications as a contraceptive and,
paradoxically, as therapy for infertility).
(Above: Magnification unknown)
(Right: Magnification unknown)

Dopamine drug crystals (polarized light micrograph)
Dopamine is a naturally occurring chemical in the brain that communicates between brain cells. It is best known for being stimulated by pleasure and reward, which it seeks with potentially addictive results. But it also plays a part in controlling movement – dysfunctional dopamine has been connected to Parkinson's Disease and attention deficit hyperactivity disorder. It has medical applications as a stimulant for low blood pressure or heart rate, sometimes in the event of cardiac arrest. (Magnification unknown)

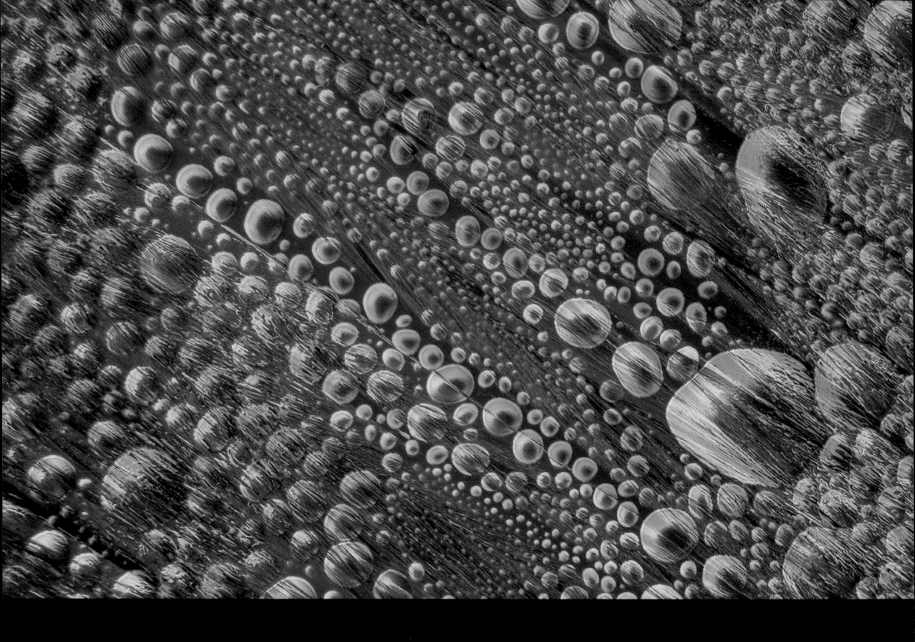

Ephedrine vapour, crystallized (polarized light micrograph)
Ephedrine was first isolated in 1881, but has been used medicinally for
about 60,000 years. It is derived from the *Ephedra* family of plants, whose

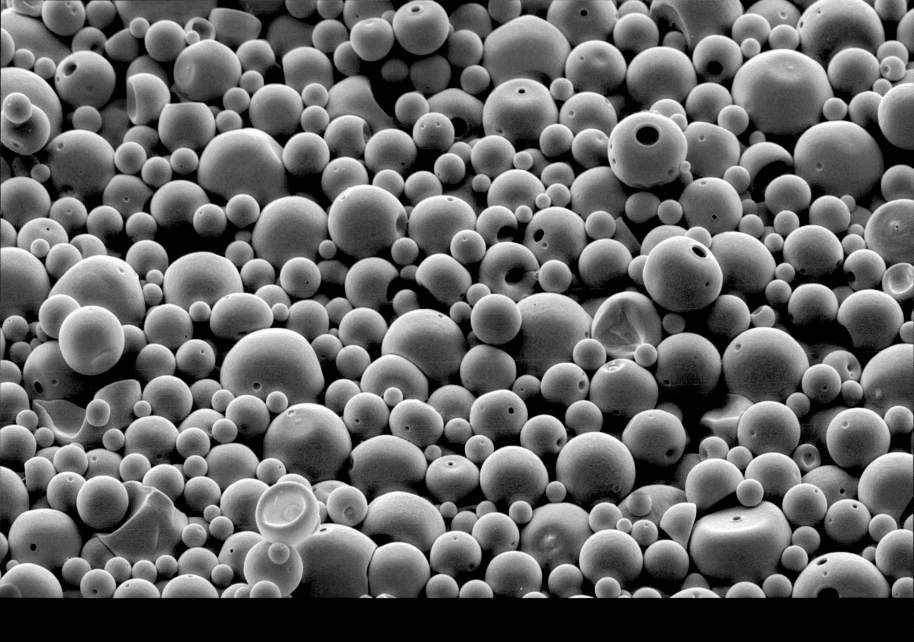

PLGA microspheres (scanning electron micrograph)
This image shows, not medicine, but a means of precisely delivering
medicine where it needs to be inside our bodies. Microspheres enclose the

Quinine (polarized light micrograph)

Quinine once put the tonic in tonic water, which was invented not as
mixer for gin but as a defence against malaria. It comes from the bark of
the Peruvian cinchona tree, which has been used against malaria since at
least the seventeenth century. Quinine was first isolated in 1820 and is
no longer recommended as a primary treatment of malaria because of
sometimes severe side effects. It does however have therapeutic benefits

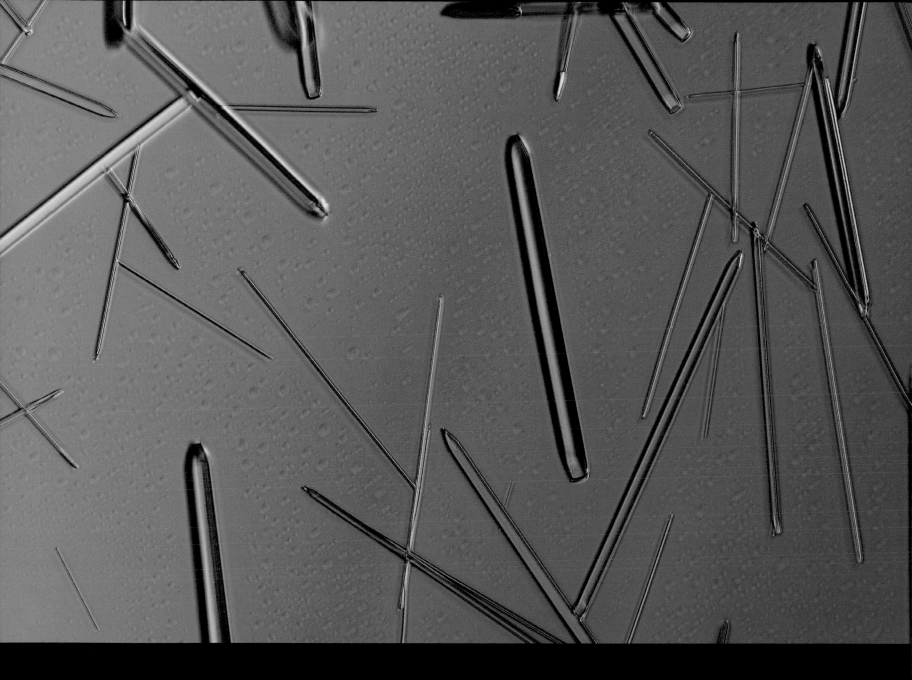

Amoxicillin crystals (polarized light micrograph)
Amoxicillin is part of the penicillin family, a versatile antibiotic that defeats
bacterial infections of the ear, nose, throat, skin (including Lyme disease)
and urinary tract (including chlamydia). When combined with clavulanic
acid it is known as co-amoxiclav and used to treat serious infections
including tuberculosis and animal bites. In conjunction with another
antibiotic, clarithromycin, it is effective against stomach ulcers caused by
Helicobacter pylori bacteria.
(Magnification x220 at 10cm wide)

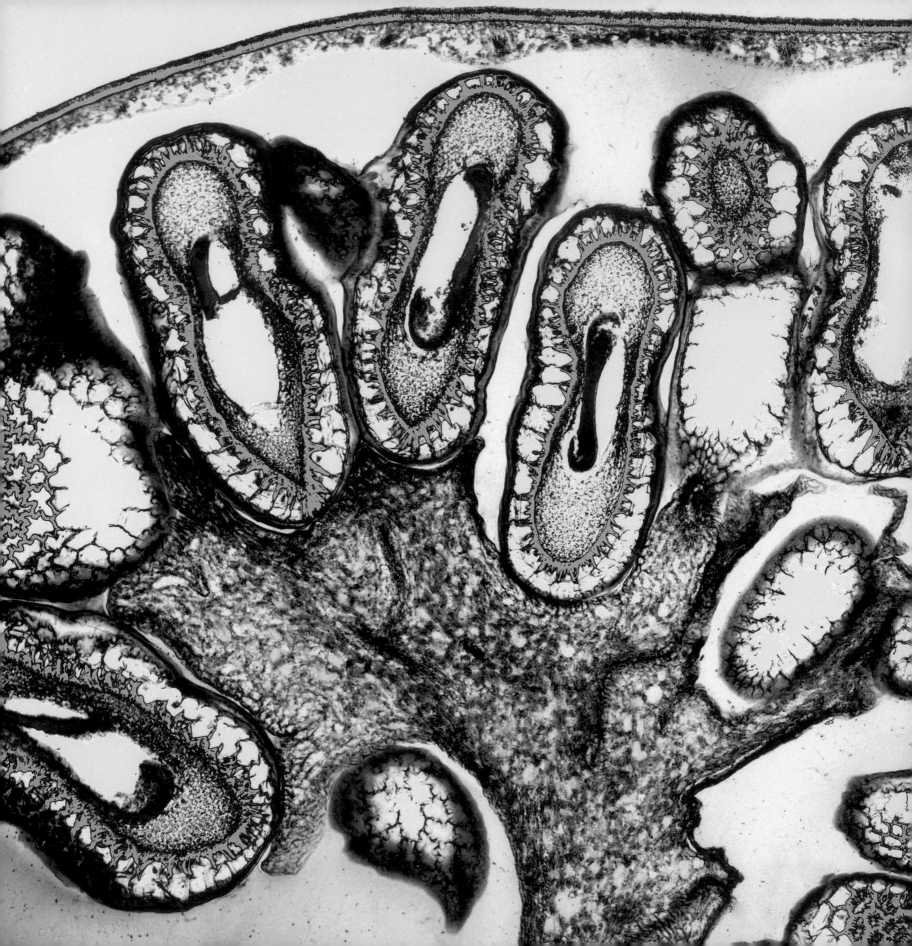

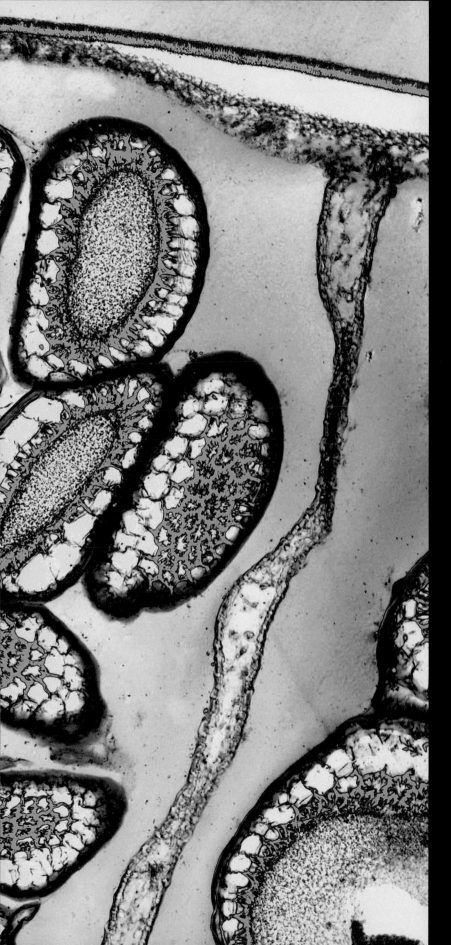

Deadly nightshade *Atropa belladonna* **(coloured light micrograph)**
This cross-section through the seed head of belladonna, the deadly
nightshade plant, shows clearly the seeds within. The nightshade family
includes tomatoes and potatoes, but the deadly strain is named for its
toxic qualities. In ancient times it was applied to the tips of arrows, and
as a fatal poison for two Roman emperors' wives. In smaller doses it is
used alongside an anaesthetic (to regualte the heartbeat) and is still an
ingredient in anti-inflammatory drugs for gastric and menstrual conditions
(Magnification unknown)

Left: Vitamin A crystals (polarized light micrograph)

Vitamin A is important for growth, the immune system and our eyesight, particularly our perception of colour and low light levels. Extreme deficiency of it can cause blindness. Vitamin A occurs in most animal products including dairy, fish (especially tuna) and meats (especially liver). But the best source for it is beta-carotene, a plant pigment found in many vegetables, including curly kale and carrots. So carrots really can help you see in the dark. (Magnification unknown)

Right: Vitamin E crystals (polarized light micrograph)

Vitamin E occurs widely, dissolved in both animal and vegetable fats. Nuts, seeds and their oils are especially rich in it. Its function in the human body is as an antioxidant: this means that it reduces wear and tear on cells, and is consequently of great interest to the anti-aging industry. Claims are made for its effectiveness in smoothing wrinkles and softening skin and hair, and in repairing burns and scars.
(Magnification x10 at 10cm wide)

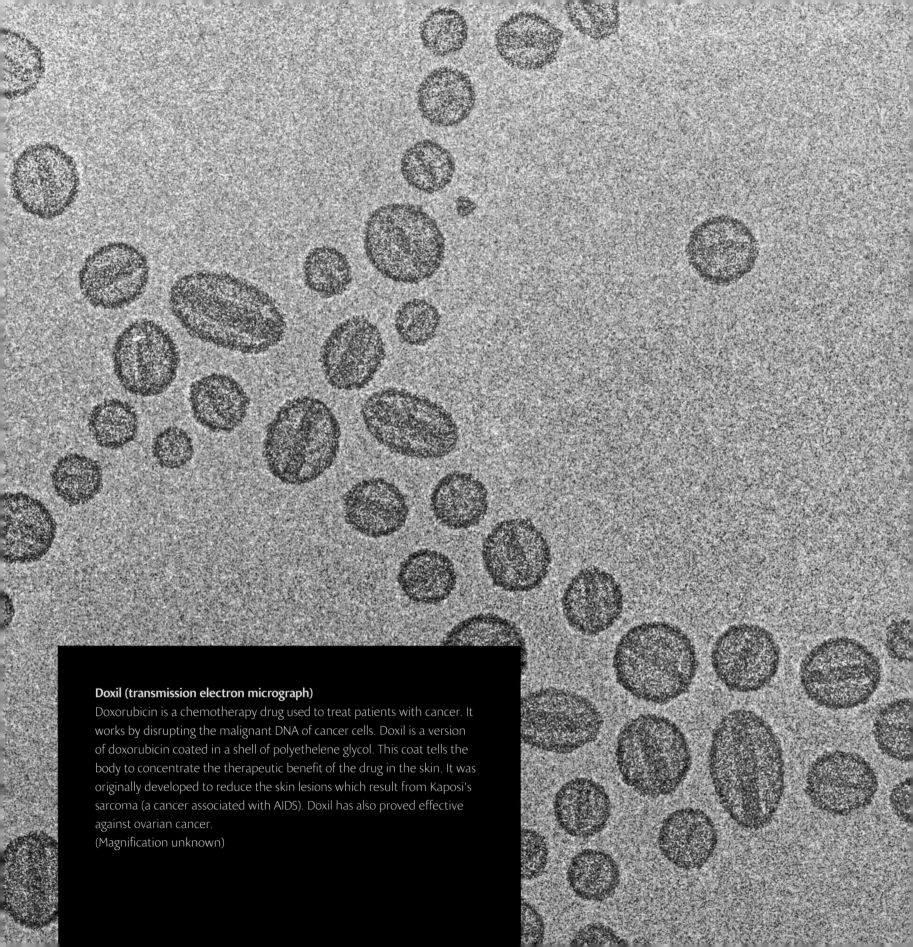

Doxil (transmission electron micrograph)
Doxorubicin is a chemotherapy drug used to treat patients with cancer. It works by disrupting the malignant DNA of cancer cells. Doxil is a version of doxorubicin coated in a shell of polyethelene glycol. This coat tells the body to concentrate the therapeutic benefit of the drug in the skin. It was originally developed to reduce the skin lesions which result from Kaposi's sarcoma (a cancer associated with AIDS). Doxil has also proved effective against ovarian cancer.
(Magnification unknown)

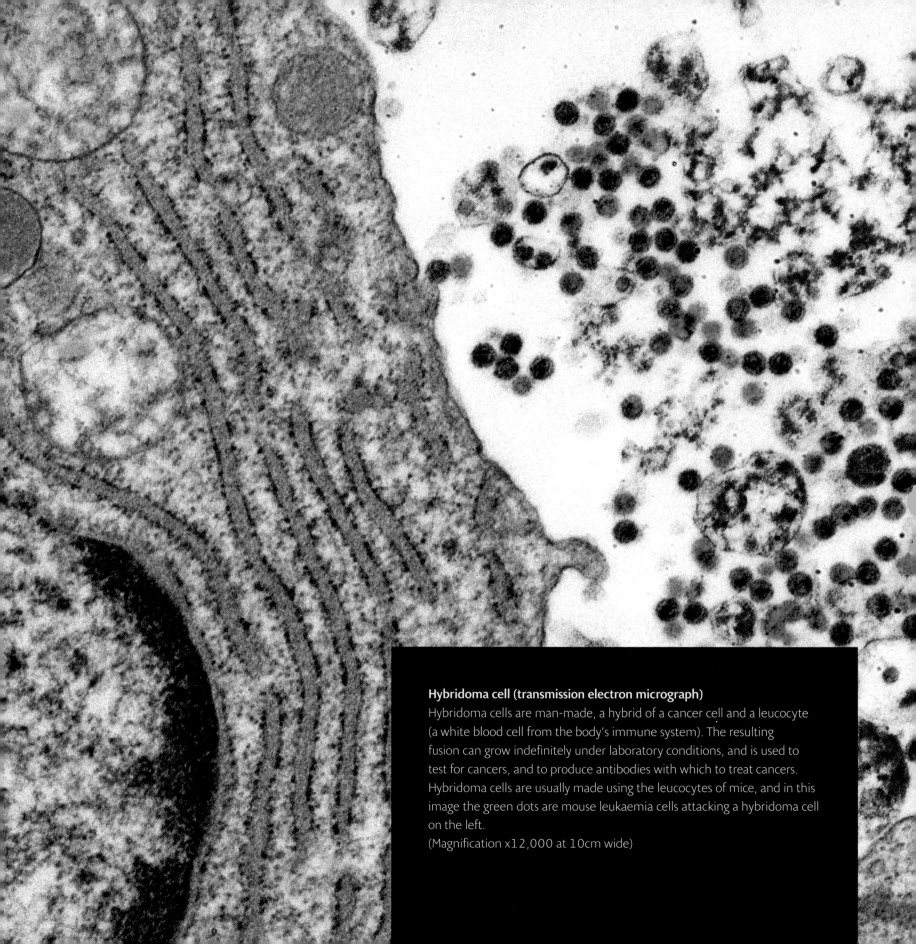

Hybridoma cell (transmission electron micrograph)
Hybridoma cells are man-made, a hybrid of a cancer cell and a leucocyte
(a white blood cell from the body's immune system). The resulting
fusion can grow indefinitely under laboratory conditions, and is used to
test for cancers, and to produce antibodies with which to treat cancers.
Hybridoma cells are usually made using the leucocytes of mice, and in this
image the green dots are mouse leukaemia cells attacking a hybridoma cell
on the left.
(Magnification x12,000 at 10cm wide)

Folic acid crystals (polarized light micrograph)
Folic acid, vitamin B9, is described as a co-enzyme, important for the development of protein and haemoglobin in our blood. A shortage of it at the embryonic stage of pregnancy can cause brain and spine defects such as spina bifida to the unborn child. Pregnant women are therefore encouraged to take extra folic acid during both conception and gestation. Some studies show that a regular intake of folic acid reduces the likelihood of strokes and heart attacks.
(Magnification x60 at 10cm wide)

Left and above: Fluoxetine drug (polarized light micrograph)

Fluoxetine hydrochloride is an anti-depressant available under several brand names of which Prozac is perhaps the most familiar. It is a selective serotonin reuptake inhibitor (SSRI); it restricts the reabsorption of serotonin, allowing more of it to remain in our brains, improving communication between brain cells. Since serotonin is responsible for feelings of happiness, SSRI's such as fluoxetine are used to treat panic, depression and obsessive-compulsive disorder.

(Left: Magnification unknown)

(Above: Magnification unknown)

Sildenafil citrate drug (polarized light micrograph)
These sharp shards are crystals of sildenafil citrate, better known by the brand name Viagra. It is used to counter male erectile dysfunction, by relaxing the muscles of the penis so that blood can flow to them and enlarge them. Sildenafil also has an application in the relief of pulmonary arterial hypertension, high blood pressure in the blood vessels of the lung. A 2007 study found that it also relieved jet lag, at least in hamsters. (magnification unknown)

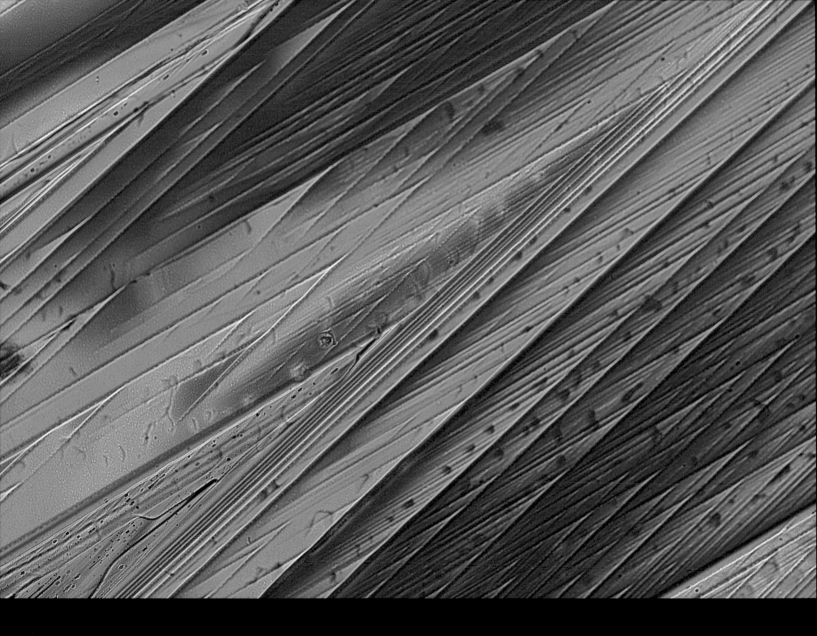

Above: Valium drug crystals (polarized light micrograph)

Valium is the brand name for the tranquiliser diazepam, first marketed after it was approved for use in 1960. It is prescribed for those suffering from insomnia, vertigo and other sources of anxiety. Diazepam is also used in the treatment of muscular spasms cause by certain conditions. Like alcohol, morphine and barbiturates it acts on the brain's system of pleasure and reward, and addiction to the drug is a possibility. Under supervision however it can ease withdrawal from other addictions. (Magnification x33 at 10cm wide)

Right: Caffeine crystals (polarized light micrograph)

Caffeine is a psychoactive drug, which means it alters perception and consciousness. Many of us rely on its stimulant effects in coffee, tea, cola or other energy drinks to enhance our performance at work. In excess, it can lead to insomnia, palpitations, disorientation, delusions and (in extreme cases) death. As a medicine it is given to premature babies and infants with breathing problems. Caffeine may also slow declines in language and cognitive function in the elderly. (Magnification unknown)

Viagra crystals (scanning electron micrograph)

Viagra, brand name for sildenafil citrate, is a common treatment of erectile dysfunction in men. In response to sexual arousal, the body releases nitric oxide, which has the effect of relaxing the muscles of the penis. This allows the blood to flow in, and the muscles swell. Viagra acts in the same way, and its use is not advised for men who may also be getting nitric oxide from elsewhere, for example heart medicines and recreational 'poppers'. (Magnification unknown)

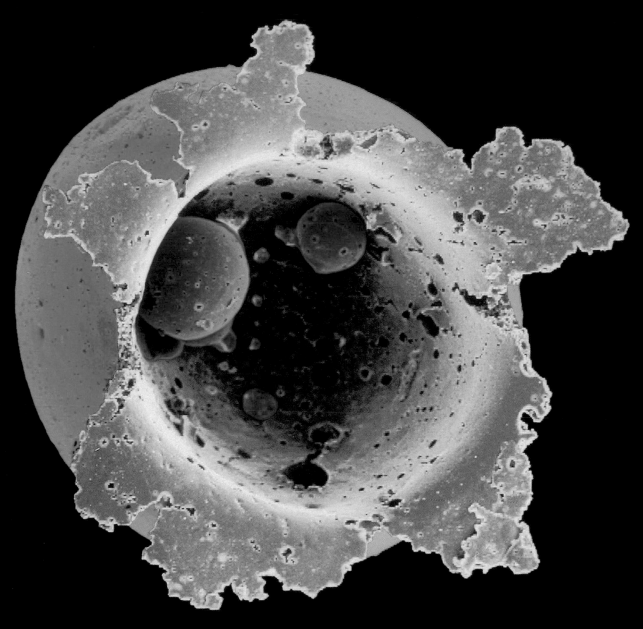

Left: Polymer sphere for drug delivery (scanning electron micrograph)
Polymer spheres are used to direct medicines to where they are required in the body. They manage this thanks to special chemical coatings, which the body recognizes and forwards to the correct location, where the coating is dissolved, the sphere bursts open and the contents are revealed. Spheres may carry drugs or (as in this image, in darker blue) other, smaller spheres to be further distributed as required. (Magnification x3000 at 10cm wide)

Right: Salbutamol sulphate crystals (scanning electron micrograph)
Salbutamol, marketed under Ventolin and other brand names, is prescribed for conditions in which breathing is made more difficult by a narrowing of airways due to inflammation. This constriction can happen after exertion, particularly in patients with asthma or chronic bronchitis. The drug is usually self-administered with an inhaler. Trials are underway to use it to treat spinal muscular atrophy, a rare inherited wasting disease caused by a protein deficiency. (Magnification unknown)

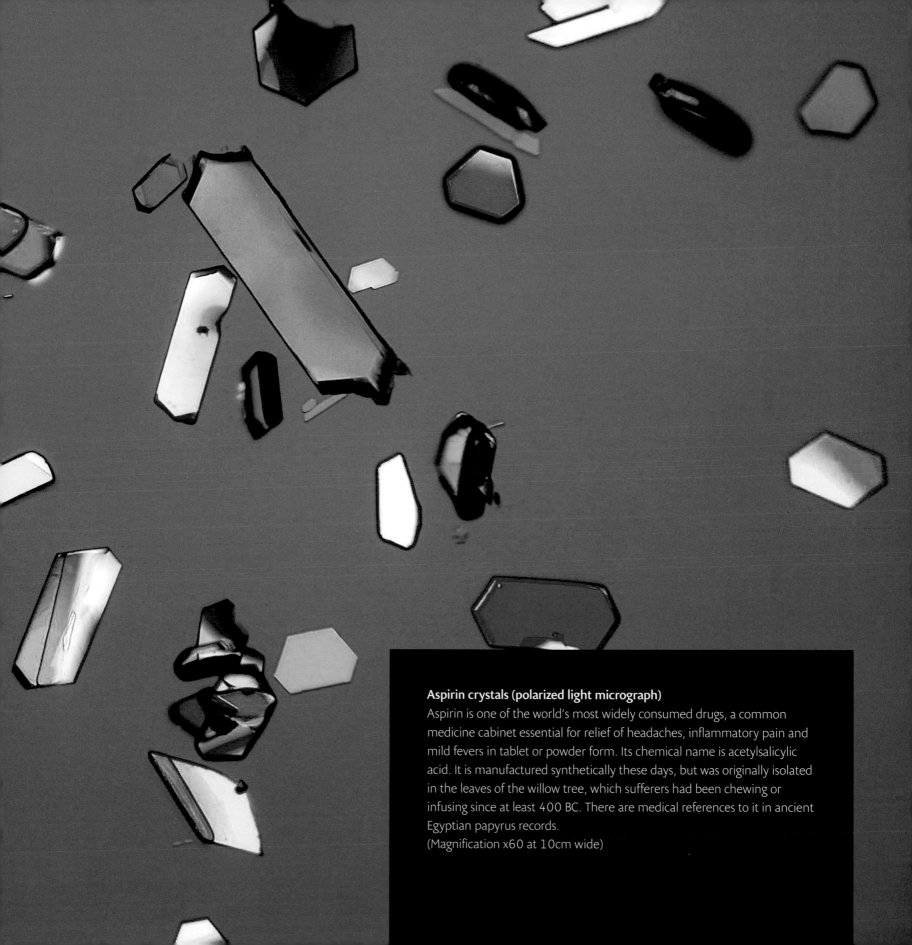

Aspirin crystals (polarized light micrograph)
Aspirin is one of the world's most widely consumed drugs, a common
medicine cabinet essential for relief of headaches, inflammatory pain and
mild fevers in tablet or powder form. Its chemical name is acetylsalicylic
acid. It is manufactured synthetically these days, but was originally isolated
in the leaves of the willow tree, which sufferers had been chewing or
infusing since at least 400 BC. There are medical references to it in ancient
Egyptian papyrus records.
(Magnification x60 at 10cm wide)

Left: Streptomycin crystals (polarized light micrograph)

In 1943, during World War II, a student at Rutgers University in New Jersey was the first to isolate streptomycin, the first antibiotic to treat tuberculosis. Its application was discovered by the US Army, which experimented with it on seriously ill soldiers after the war, persevering after early patients died or went blind. The student's professor was later awarded the Nobel Prize for his pupil's achievement.
(Magnification x33 at 3.5cm wide)

Above: Crystals of the drug Vidaza (coloured light micrograph)

Vidaza is a brand name for azacitidine, a chemotherapy drug which treats blood disorders known as myelodysplastic syndromes (MDSs). MDSs restrict the development of immature blood cells in bone marrow, resulting in low counts and abnormalities of platelets and red and white blood cells. Blood transfusions are a standard treatment, but the introduction of Vidaza in 2004 made them less necessary. Vidaza stimulates normal blood cell production and also destroys malignant blood cells.

Left: Aspirin (polarized light micrograph)
Aspirin is a common non-steroid anti-inflammatory drug (NSAID) used to treat everyday aches and pains. Because it also acts to thin the blood it can prevent heart attacks, and it may decrease the possibility of developing colorectal cancer. The name was in 1897 the brand of the drug's creators, Bayer. But although Bayer still holds some rights to the name, aspirin (with no capital letter) has become the generic name for this sort of painkiller in many countries.
(Magnification unknown)

Right: Zorvirax crystal (transmitted light micrograph)
Zovirax is a brand name for the antiviral drug acyclovir. Discovered in 1977, acyclovir is used for the prevention and treatment of several strains of the herpes virus. These include genital herpes, cold sores, chickenpox and chronic eye herpes conditions. A Caribbean sponge was the original source of the drug, which is able to target viruses selectively and without collateral damage to uninfected cells.
(Magnification unknown)

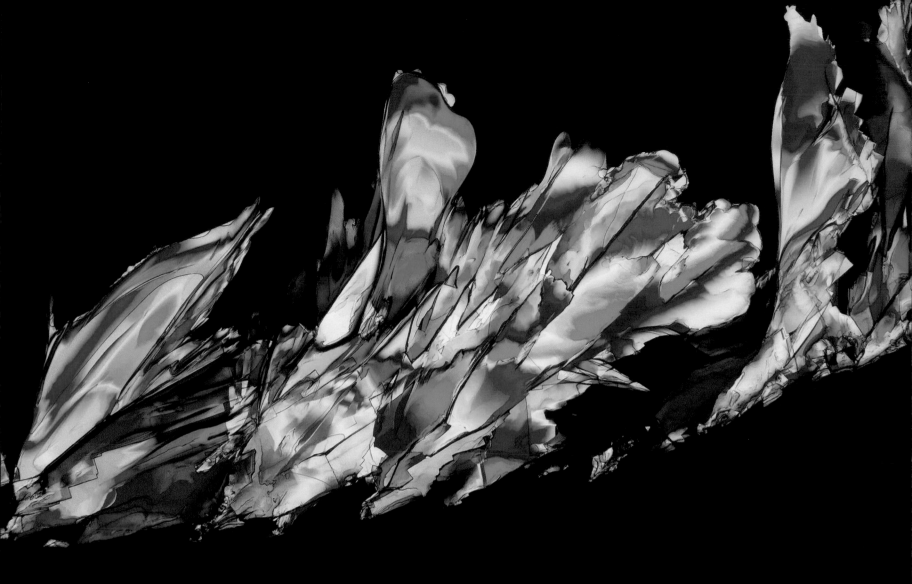

Above: Dopamine drug crystals (polarized light micrograph)
Dopamine is a naturally occurring neurotransmitter in the brain, where it helps to regulate our centres of pleasure and reward. Unusually, the drug was synthesized (in 1910) before it was discovered in its natural form in the human brain (in 1957). Dopamine has been found in almost all animals, including bacteria. Many plants also synthesize dopamine, notably the banana; but plant dopamine cannot cross the blood-brain barrier – so eating bananas will not necessarily make you happy. (Magnification unknown)

Right: *Cannabis sativa* (scanning electron micrograph)
These artificially coloured bumps on the surface of a cannabis plant are glands called trichomes, which secrete the cannabis resin known to recreational drug users as hashish. Marijuana is derived from flowers and leaves of the plant. Kief (from the Arabic word for pleasure) is a powder residue from cannabis flowers, leaves and trichomes. Medical cannabis relieves chronic pain and muscle spasms, including the tics of Tourette's syndrome. It also reduces the nauseous side effects of chemotherapy. (Magnification x35 at 10cm wide)

Ephedrine drug crystals (polarized light micrograph)
Ephedrine is used medically to raise blood pressure, and often prescribed
to open the airways of asthma sufferers and others. It also promotes
weight loss, particularly when combined with caffeine (found naturally in
coffee and elsewhere) and theophylline (a component of cocoa beans).
(Magnification unknown)

Pantothenic acid crystals (polarized light micrograph)

This artful arrangement of crystals is derived from an image of pantothenic acid, vitamin B5. The vitamin is a useful boost for those who need it – the very young and very old, and pregnant and post-natal women. It helps to process fats, carbohydrates and proteins. The name comes from ancient Greek, meaning 'from everywhere', and pantothenic acid is found in almost every foodstuff. Liver, kidney, egg yolk and peanuts are among the richest sources.

(Magnification unknown)

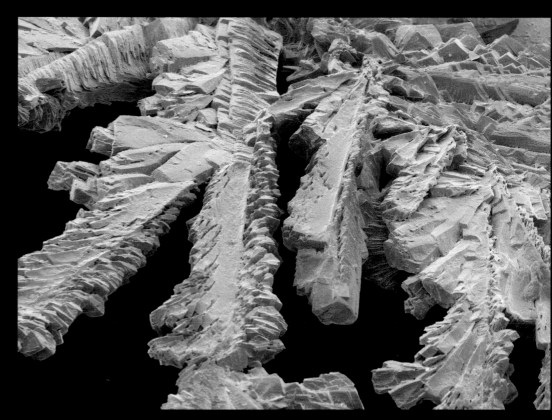

Above: Antihistamine drug crystals (scanning electron micrograph)
Histamines are released by the body's immune system as a defence
against invading allergens such as pollen. They release fluids into body
tissue, typically causing streaming eyes and a runny nose. But the nose
can also become blocked and itchy as the tissue swells and inflames.
Antihistamines block this defensive reaction Their use against hayfever
is well known, but they are also given to patients with potentially fatal
allergic reactions known as anaphylactic shock.
(Magnification unknown)

Left: Ketamine crystals (polarized light micrograph)
Ketamine is used for sedation and pain relief, often in emergency
situations in hospital or on the battlefield of war. It is less likely than some
anaesthetics to disrupt reflex actions such as heartbeat and breathing.
Because ketamine induces a state of trance, recreational drug users have
experimented with it, but its unsupervised non-medical use has resulted
in many careless deaths. It causes detached, sometimes hallucinatory
sensations; and some users have for example accidentally drowned or
poisoned themselves.
(Magnification unknown)

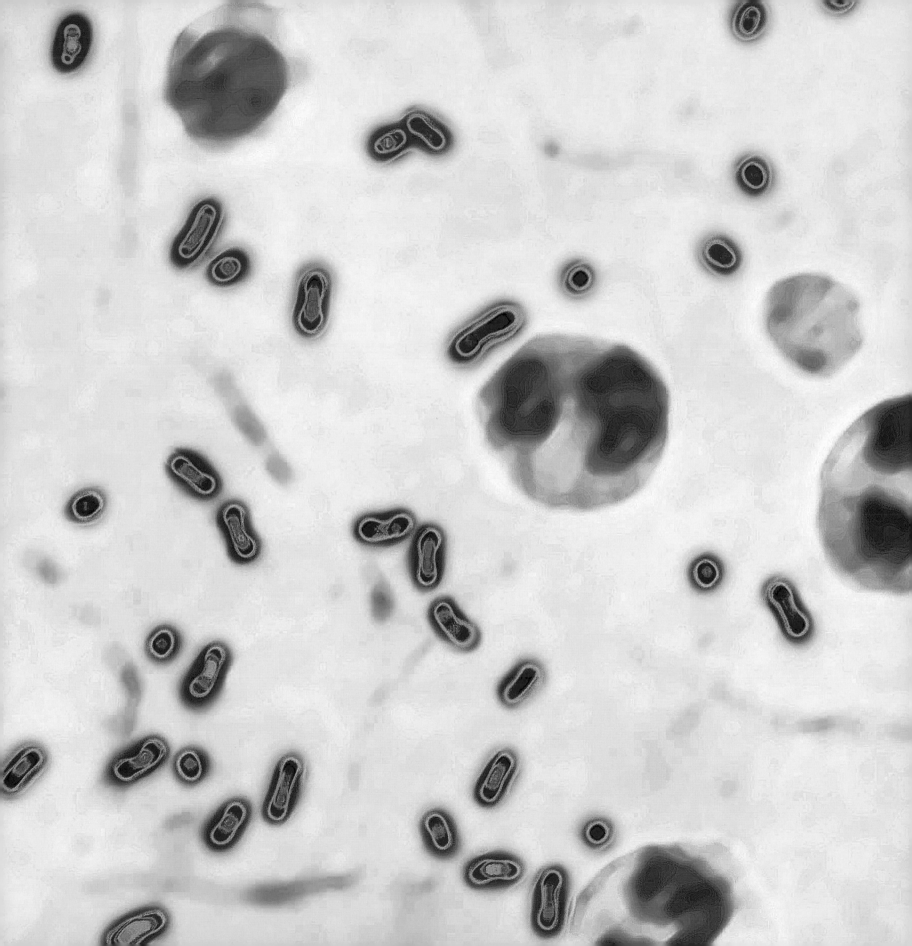

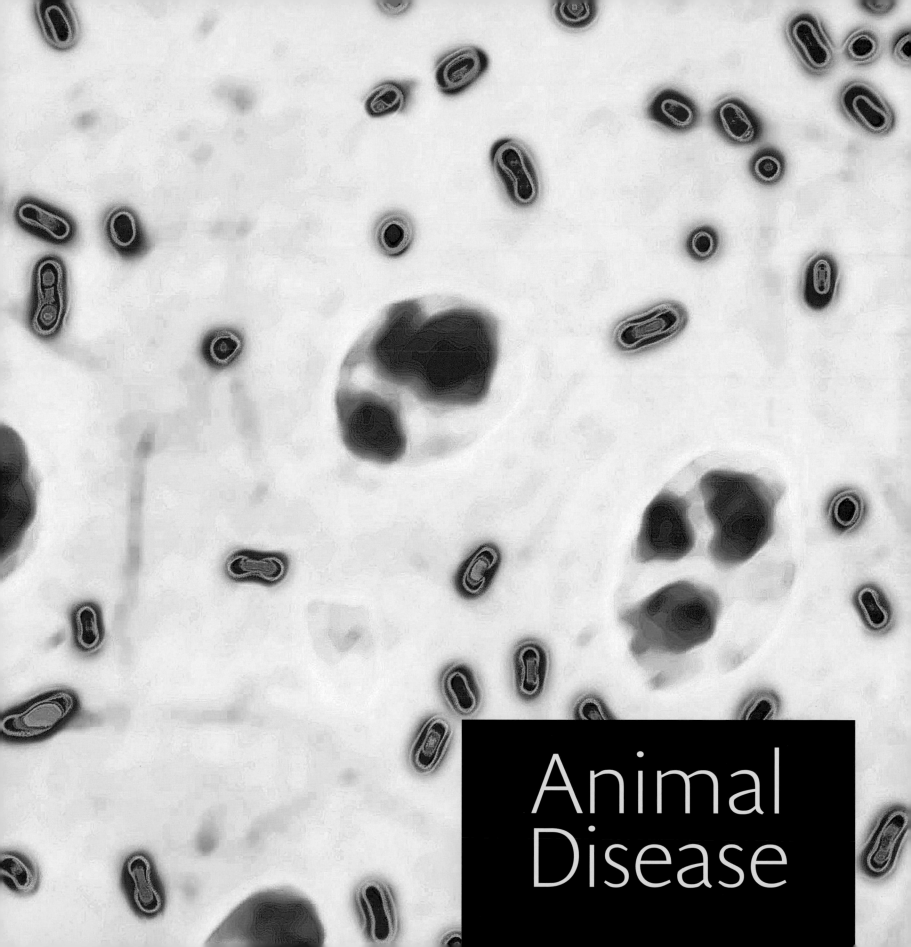

Animal
Disease

The bubonic plague, also known as the Black Death, is estimated to have killed around a third of the world's population in a single outbreak in the fourteenth century. The little blue ovals in this image were responsible – *Yersinia pestis* bacteria. The bacteria's hosts were the fleas of rats, which spread the disease around the globe by following armies, trade routes and (ironically) the paths of those fleeing the plague itself.
(Magnification unknown)

Right: *Yersinia pestis* **bacteria (scanning electron micrograph)**
The symptoms of bubonic plague start with flu-like symptoms and end with erupting boils on the skin where a rat flea has introduced the *Yersinia pestis* bacteria. This enlarged image shows the spines of an oriental rat flea (here coloured purple) infested with *Yersinia pestis* (yellow grains of rice). The bacteria spreads not one but three types of plague – bubonic, pneumonic and septicaemic. Thousands of cases are still recorded every year, but modern medicine has greatly improved the prognosis for patients.
(Magnification unknown)

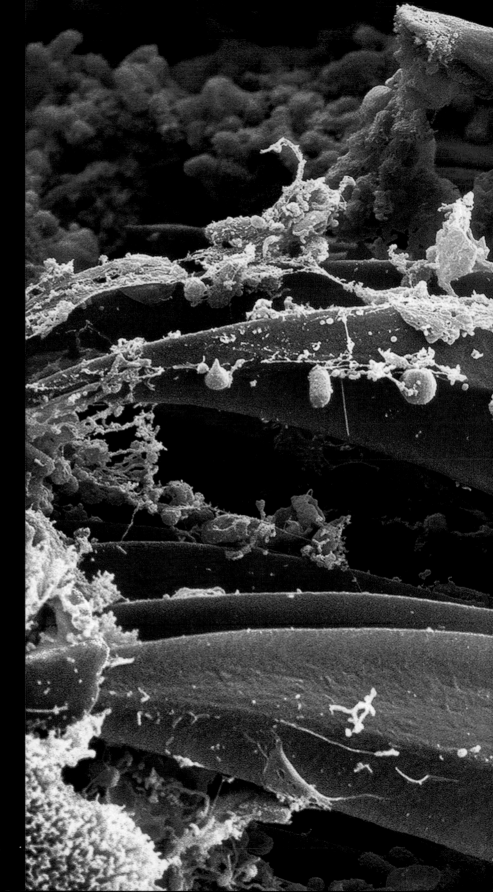

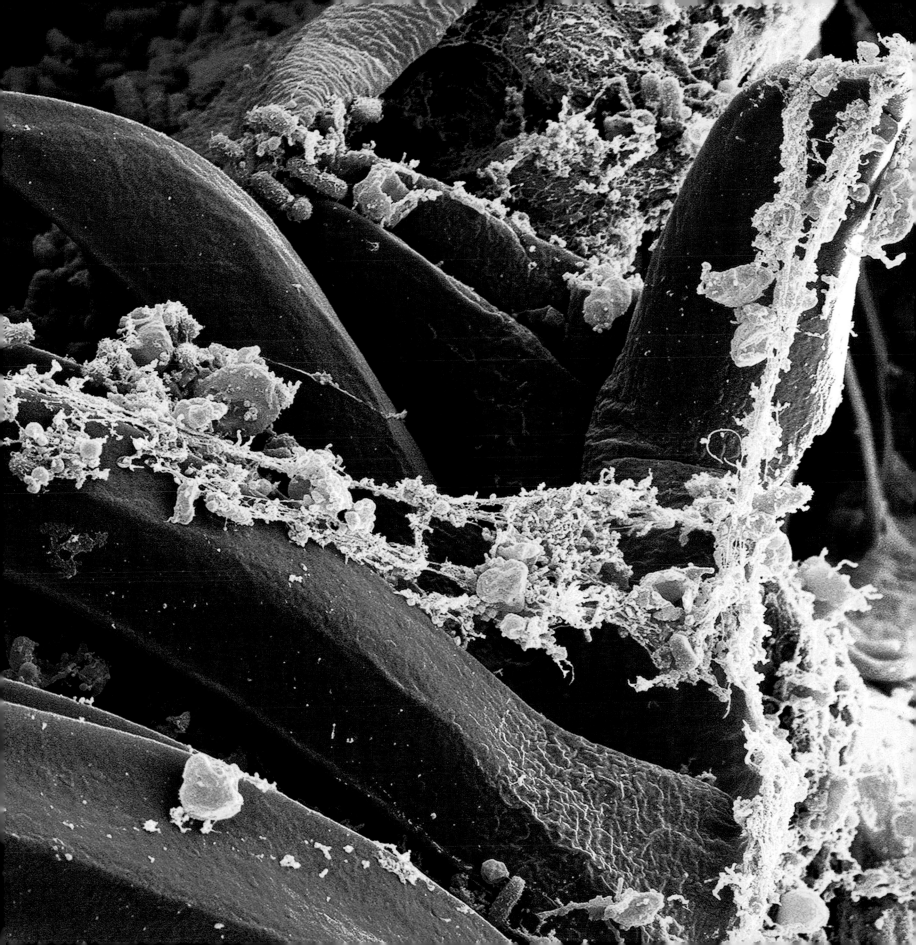

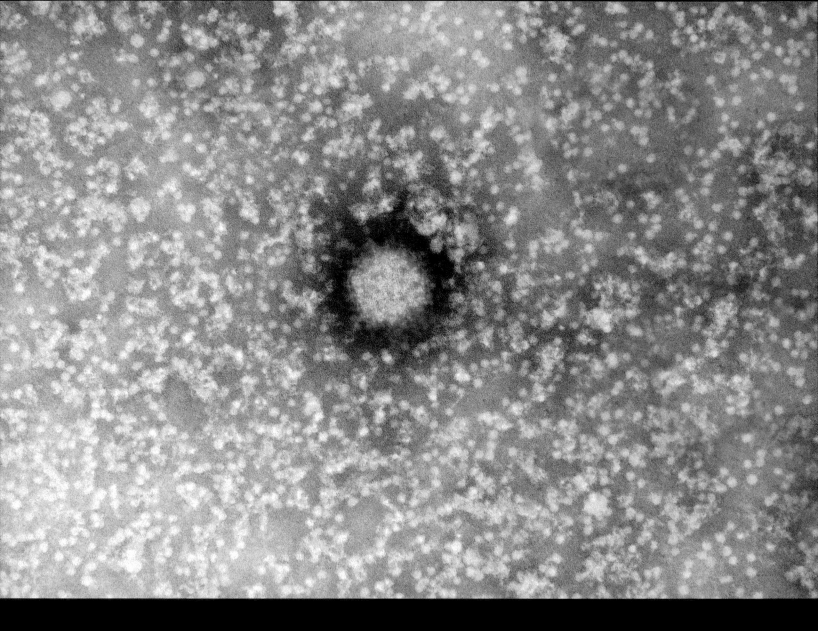

Above: Schmallenberg virus (transmission electron micrograph)
This orange sun in a green starry sky is a single particle of the
Schmallenberg virus, a disease of farm livestock named after the health
resort in northern Germany where it was first identified in 2011. Since
then it been reported in more than fifteen further European countries.
The virus, carried by midges, causes stillbirth and congenital deformities in
stock. It is hoped that a vaccine for a similar virus will also prove effective

Right: *Yersinia pestis* (fluorescent light micrograph)
A fluorescent antibody has been introduced to highlight this swarm of
Yersinia pestis, the bacteria that passes to humans through the bite of
infected rat fleas. The bacteria have resulted in several historic pandemic
outbreaks of bubonic plague around the world. In the modern age,
however, very prompt treatment with antibiotics can avert death in
most cases.

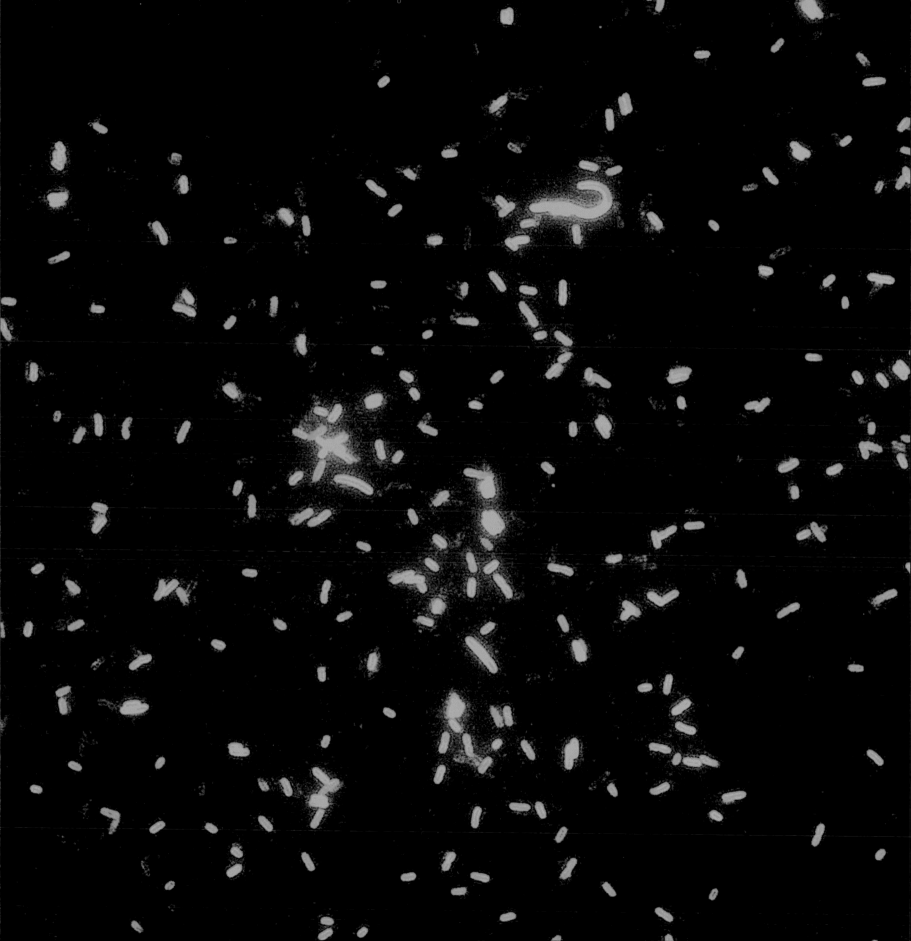

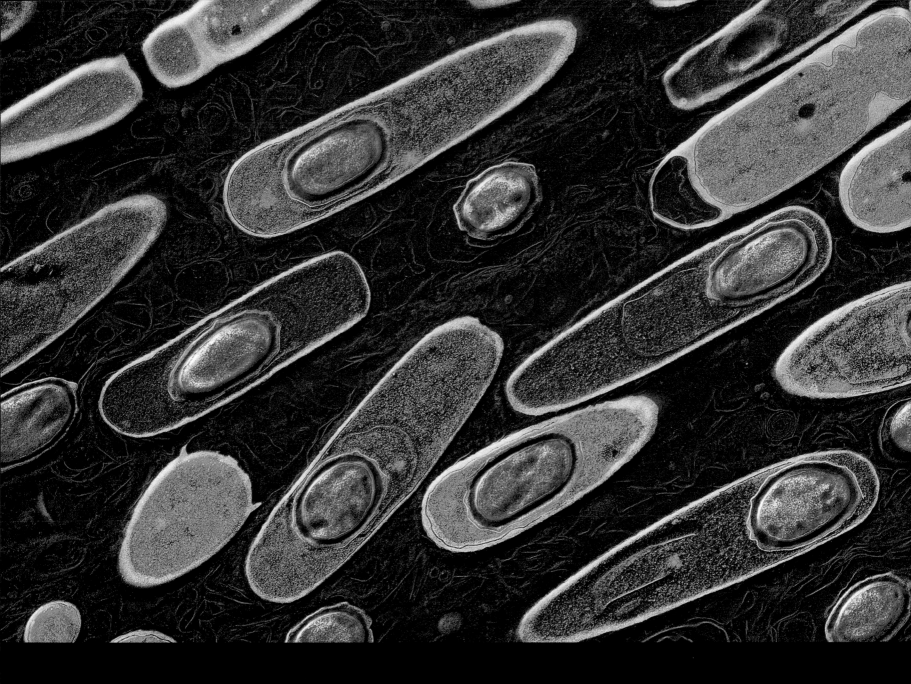

Above and right: Anthrax bacteria (transmission electron micrograph)
In these two images, the red or pink circles are spores of *Bacillus anthracis*, which spread anthrax. Some of the spores are encased in an outer shells, some are free – a spore can survive within its shell in a dormant state for many years until conditions are favourable for it to reproduce. This highly infectious disease is usually caught from animals, either by inhaling the spores, by consuming infected meat, or by direct contact with the skin. (Above: Magnification, x9,300 at 10cm high)

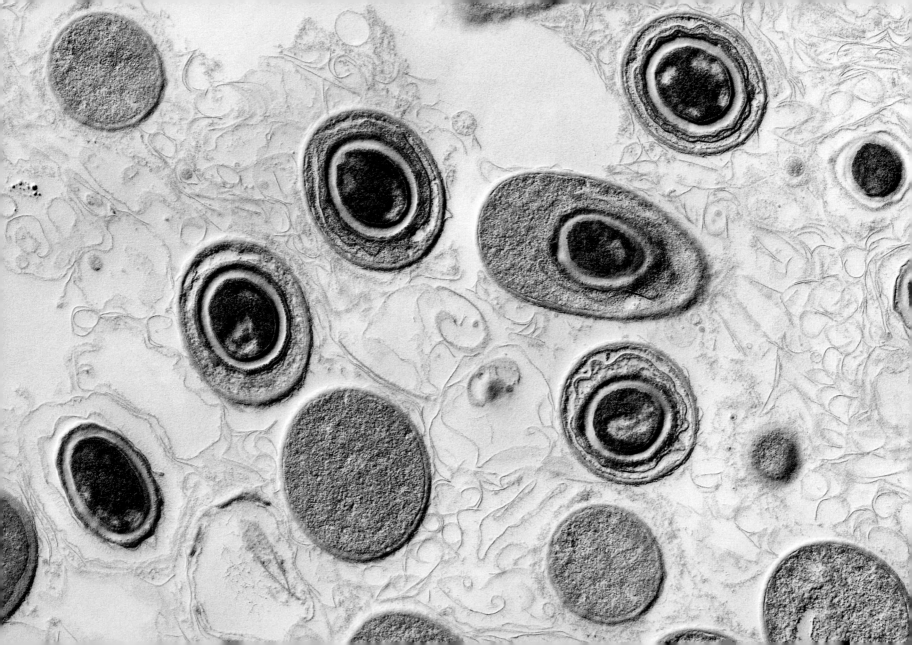

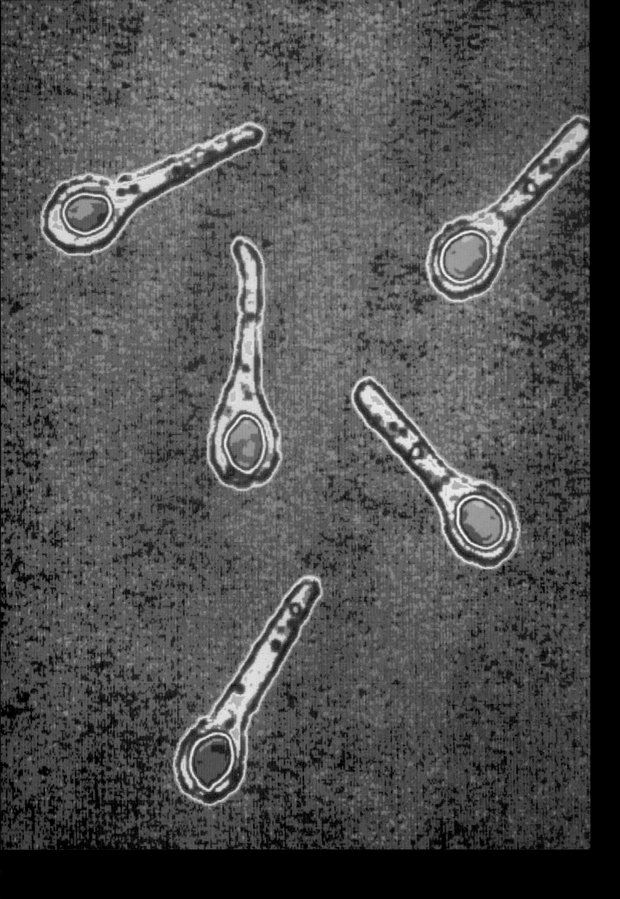

**Left: *Clostridium tetani* bacteria
(transmission electron micrograph)**
These bacteria live in the soil and in the gut of animals. When they enter the human body through an open wound, they cause tetanus, of which symptoms include painful muscular spasms and breathing difficulties. The toxin produced by the bacteria is, by weight, one of the most powerful known: a dose of 2.5 nanograms per kilogram of bodyweight is lethal. A vaccine developed in 1924 saved the lives of many wounded men in World War II. (Magnification unknown)

**Right: Cell infected by toxoplasma
(transmission electron micrograph)**
In this picture, the small pink discs within the larger green one are parasites in a human cell. The parasite, *Toxoplasma gondii*, is found in all warm-blooded animals, although it can only reproduce in wild and domestic cats. It causes toxoplasmosis, a disease that produces few or no symptoms in healthy adults. In those with weakened immune systems however it can induce seizures and other difficulties with movement. Mothers may pass it to their unborn children.
(Magnification x4170 at 10cm wide)

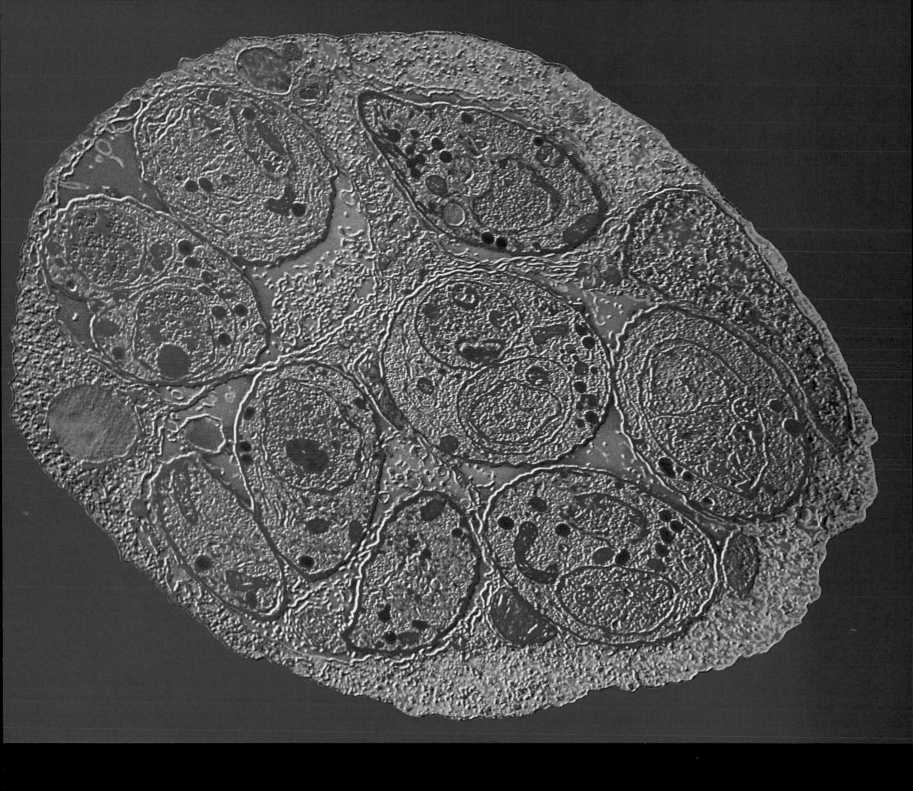

Left: Lyme disease bacteria on sheep tick (scanning electron micrograph)
In this image the serrated edges are the elongated jaws of a sheep tick and the tiny red rods are *Borrelia burgdorferi*, responsible for Lyme disease. Ticks feed on the blood of birds, reptiles and mammals, and the disease is spread when it bites through the skin of its victim, usually after the tick has been attached for a day or two. Lyme disease has a range of unpleasant symptoms, which can recur months after the initial infection.
(Magnification x550 at 6cm wide)

Right: Dog tapeworm rostellum (coloured light micrograph)
This peculiar blue star is the beak or rostellum of a dog tapeworm. The worm uses the circle of hooks on the end of it to latch on to the intestine of its host, so that it is not forcibly detached by incoming food or outgoing faeces. Each hook is about 240 micrometers or one hundredth of an inch in length. By contrast the tapeworms themselves may grow up to two metres long.
(Magnification unknown)

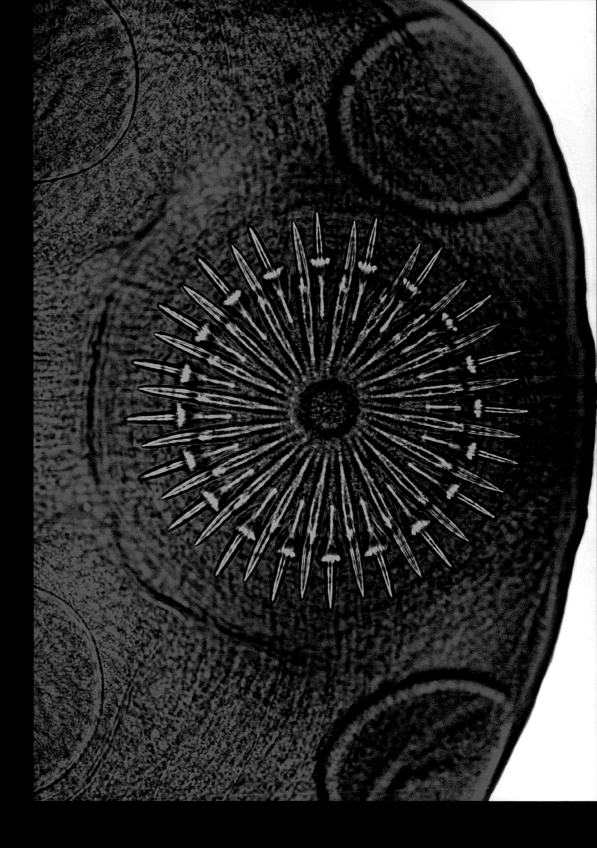

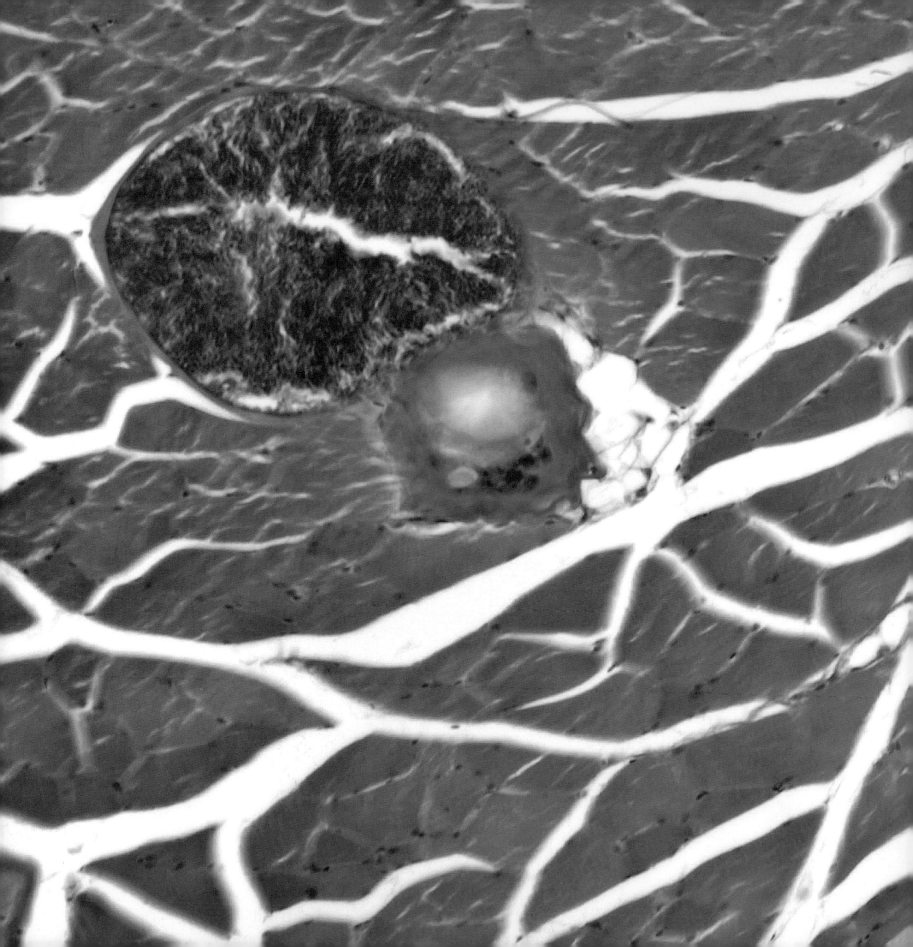

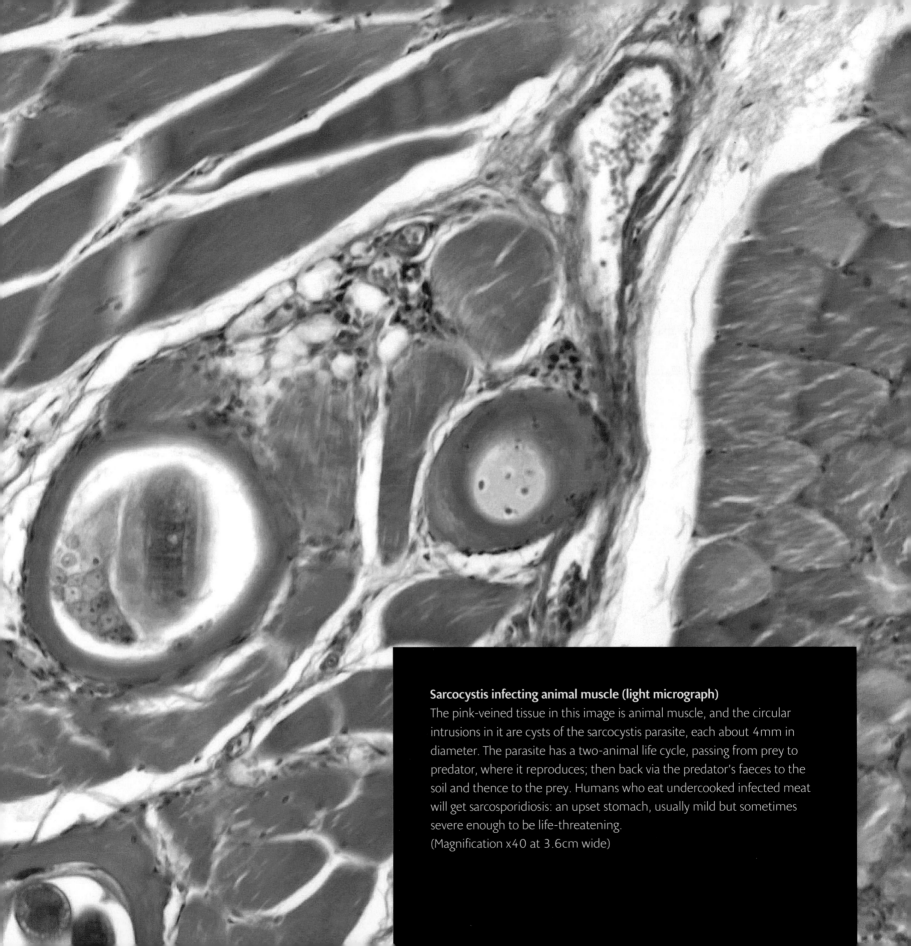

Sarcocystis infecting animal muscle (light micrograph)
The pink-veined tissue in this image is animal muscle, and the circular intrusions in it are cysts of the sarcocystis parasite, each about 4mm in diameter. The parasite has a two-animal life cycle, passing from prey to predator, where it reproduces; then back via the predator's faeces to the soil and thence to the prey. Humans who eat undercooked infected meat will get sarcosporidiosis: an upset stomach, usually mild but sometimes severe enough to be life-threatening.
(Magnification x40 at 3.6cm wide)

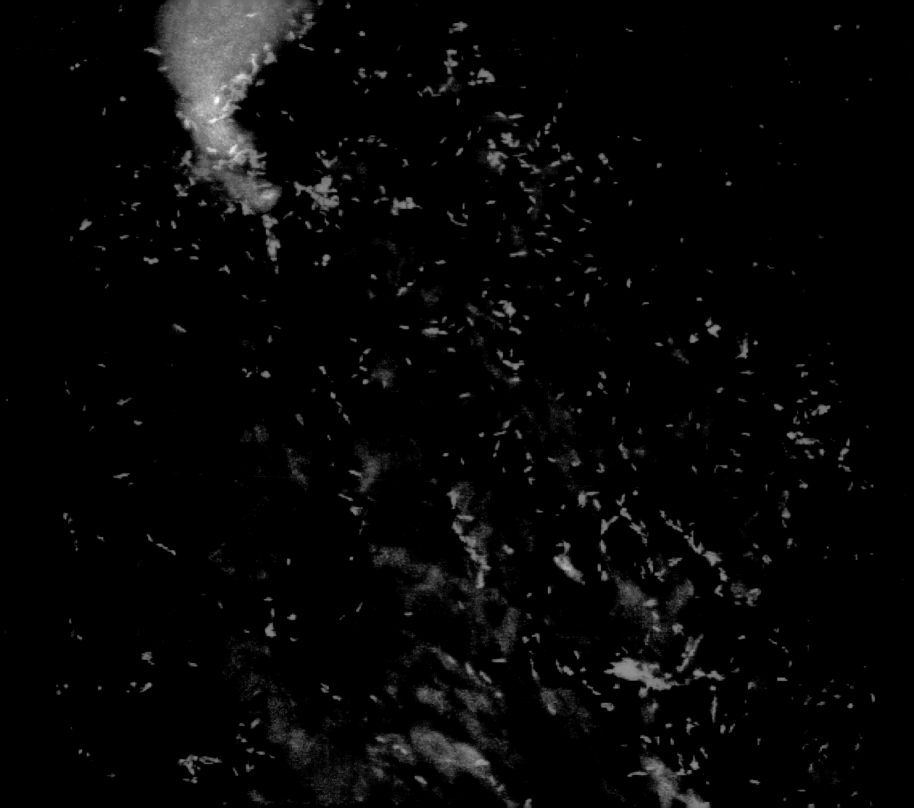

Left: *Campylobacter* bacteria on chicken skin (confocal light micrograph)

Campylobacter bacteria, the green flecks in this image, most commonly live in poultry. They may infect humans who eat raw or contaminated chicken. The usual symptoms of campylobacteriosis, the resulting illness, are up to a week of painful stomach cramps and bloody diarrhoea. *Campylobacter* prolongs its life with a toxin that prevents infected human cells from dividing, a process which would normally activate the body's immune system. Drinking plenty of fluids flushes the bacteria out.
(Magnification unknown)

Right: *Francisella tularensis* bacteria (transmission electron micrograph)

These bacteria live in rodents such as squirrels and rabbits but may infect humans with tularaemia through inhalation, broken skin, or most commonly by a tick bite. Hunters and farmworkers are most at risk. As few as ten *Francisella* bacteria may cause infection, and tularaemia has been classified along with ebola, anthrax and bubonic plague as potential threats in biological warfare. Rarely fatal, it causes incapacitating fever and pneumonia; there is no publicly available vaccine.
(Magnification x58,000 at 6cm wide)

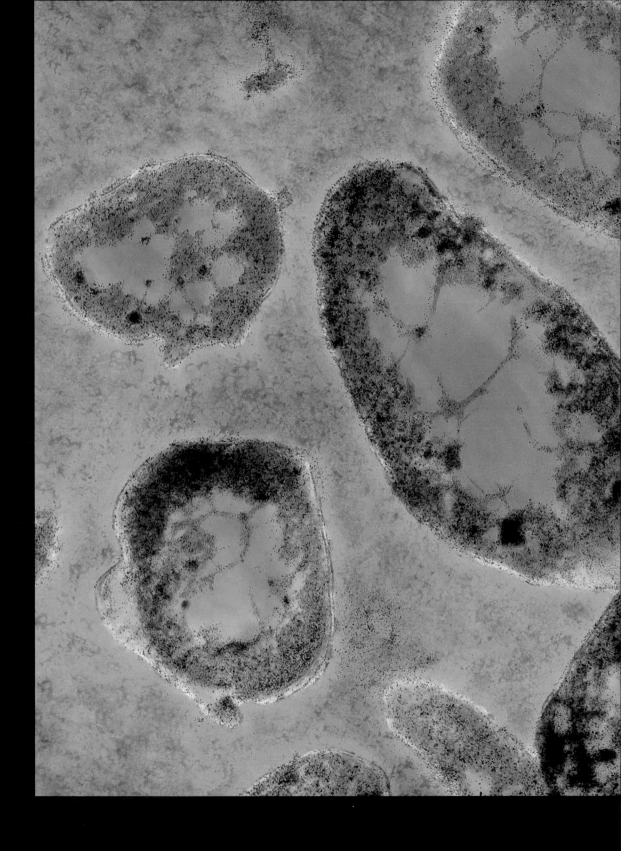

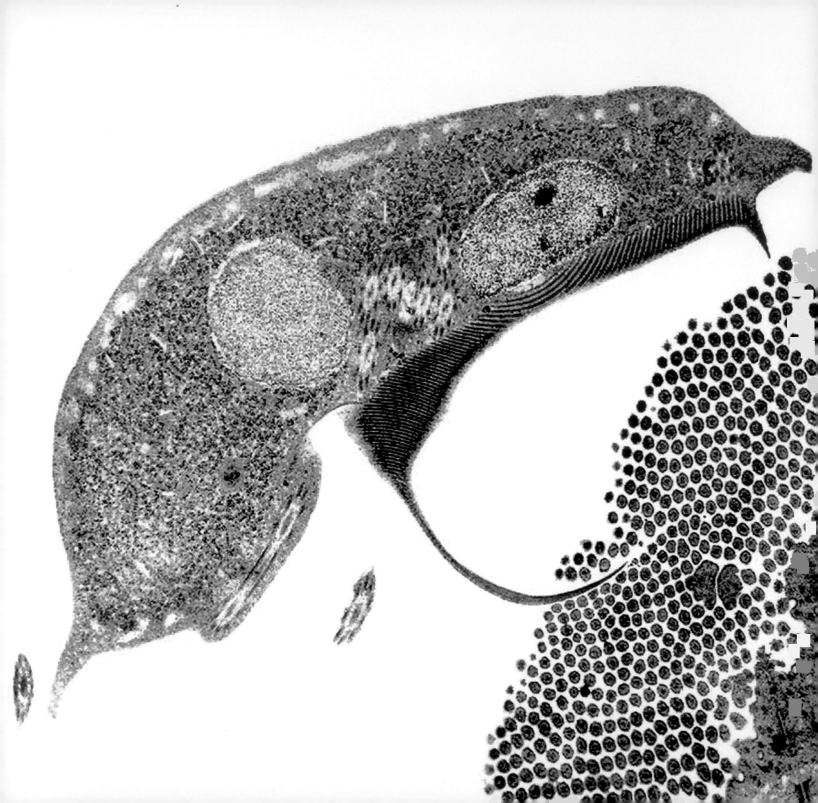

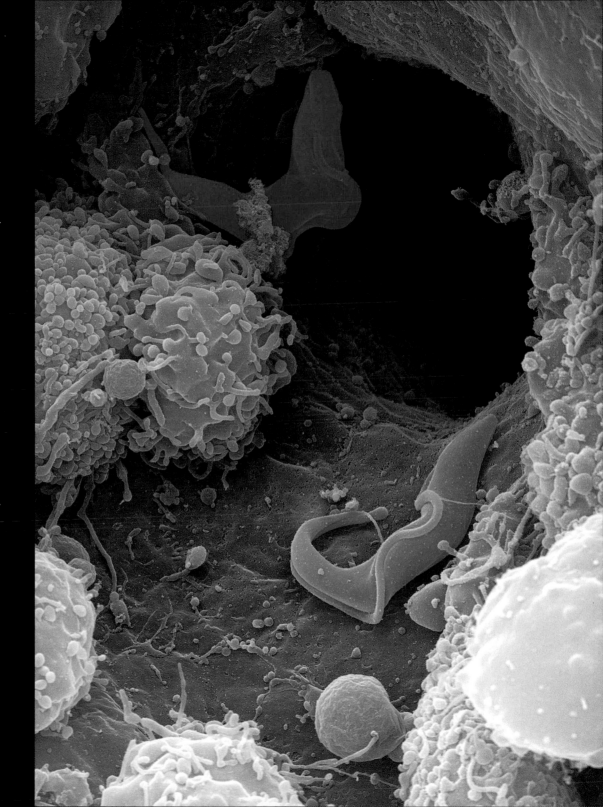

Left: Giardia protozoan parasite (transmission electron micrograph)

In this cross-section, the small green circles are microvilli, the hairlike nutrition-absorbing pimples that line the intestines. The blue creature is a giardia parasite. The hook-like limbs are its flagellae, which it uses for propulsion. It attaches itself to the intestine with a large suction pad, of which the concentric circles can be partly seen on its body to the lower right. Humans acquire it from contaminated food and water and develop violent stomach upsets.
(Magnification x7,300 at 10cm wide)

Right: Sleeping sickness parasites (scanning electron micrograph)

The figures coloured blue in this landscape of white blood cells are *Trypanosoma brucei* parasites, responsible for so-called sleeping sickness, trypanosomiasis. The bloodstream is infected with the parasite by the bite of the tsetse fly, which is common throughout sub-Saharan Africa. Symptoms begin three weeks later with joint pain and fever. After a few weeks parasites spread to the central nervous system, leading to disrupted sleep, and physical and mental confusion. Untreated, the disease is always fatal.
(Magnification x3,500 at 10cm wide)

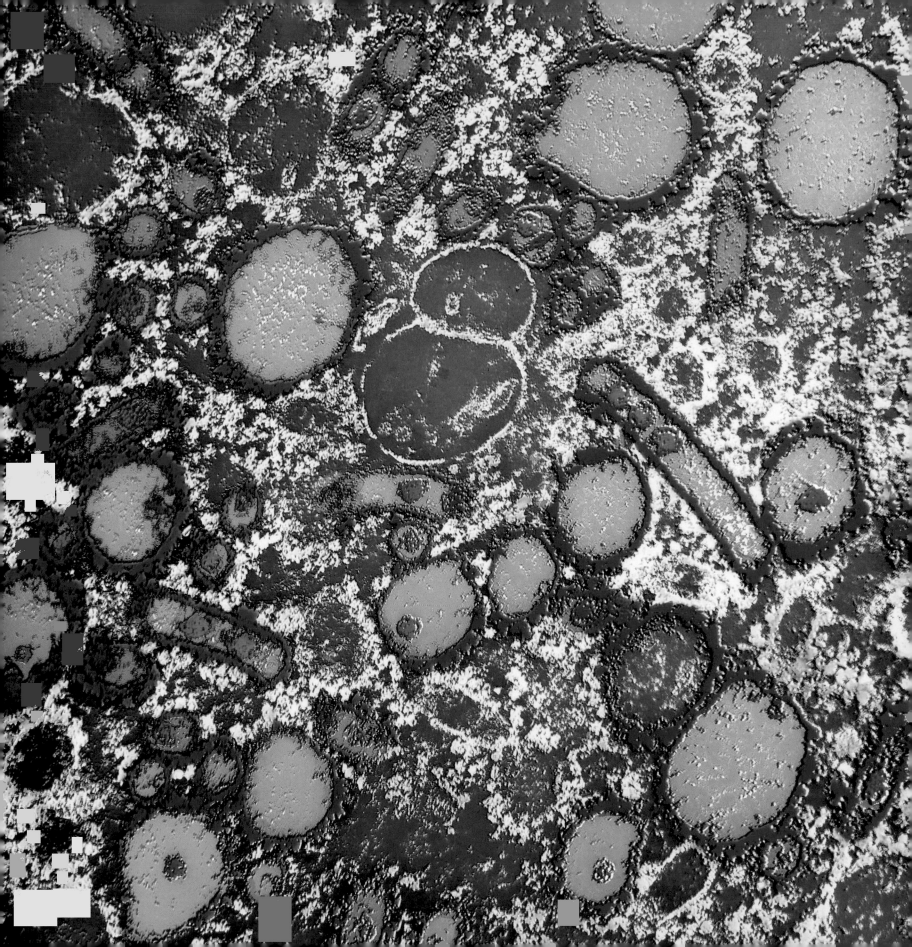

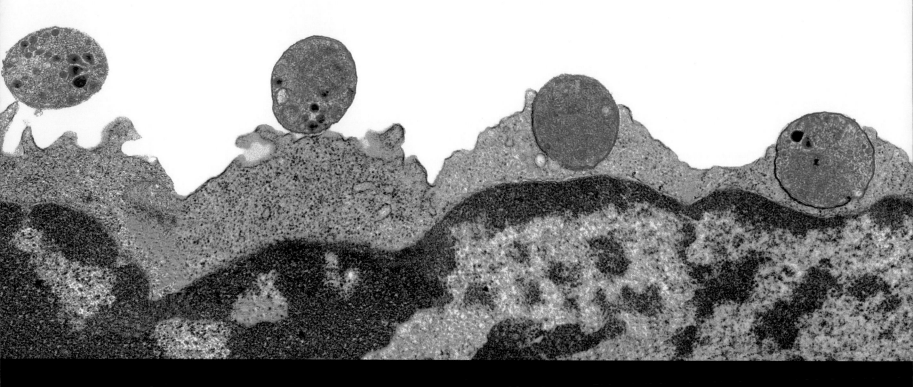

**Left: Bovine viral diarrhoea virus in cell
(transmission electron micrograph)**

This image shows a cell from a bull's testicle, infected with bovine
viral diarrhoea virus (BVDV). BVDV is a worldwide disease of cattle that
lowers milk yield, fertility and resistance to disease. Most overcome the
disease after a few weeks. In animals with an incomplete immune
system, however, the virus is accepted as normal by the body and
becomes persistent – that is, permanent. Persistently infected cattle
transmit a thousand times more viral matter than those with only a
passing infection.
(Magnification x20,000 at 6cm tall)

**Above: *Theileria parva* infecting bovine lymphocyte
(transmission electron micrograph)**

Two members of the *Theileria* family of parasites are responsible for
significant cattle diseases. This composite image shows several steps in
the infection process by *Theileria parva*, as it approaches, attaches to, and
invades a cow's white blood cell. *T. parva* causes East Coast fever, one of
the deadliest cattle infections in the African continent. *Theileria* parasites
are all spread by tick bites. Another member, *Theileria microti*, causes
human theileriosis, which has malaria-like symptoms.
(Magnification unknown)

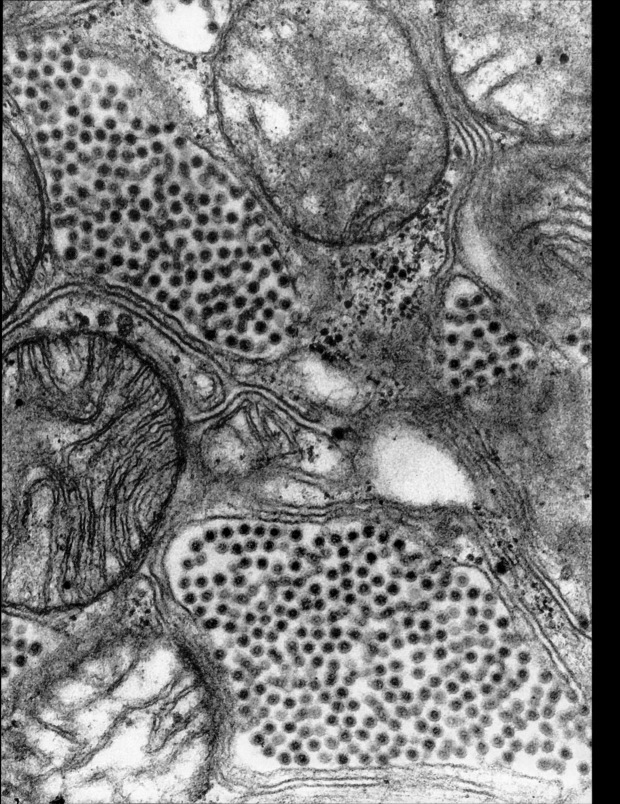

Left: Eastern equine encephalitis virus (transmission electron micrograph)

The green dots here are particles of Eastern equine encephalitis (EEE) virus, in the saliva gland of a mosquito. Mosquitoes carry the virus not only to horses but also to birds and humans in the eastern states of North, Central and South America. Symptoms include seizures and sensitivity to sound or light. There is no known cure: it is fatal in about 30 per cent of human and 80 per cent of equine cases, although there is a preventative vaccine for horses. (Magnification unknown)

Right: *Enterocytozoon bieneusi* parasite (transmission electron micrograph)

Enterocytozoon bieneusi attacks the lining of the intestine, and those with compromised immune systems are vulnerable. It was first found in 1985 in an AIDS patient. The parasite invades the cells of the intestinal wall, where it reproduces asexually before bursting out of the infected cell. In this image a mature reproduced spore of the parasite is bursting out of its host cell. The infection causes diarrhoea, returning the parasite to the environment in faeces. (Magnification x2,222 at 6cm wide)

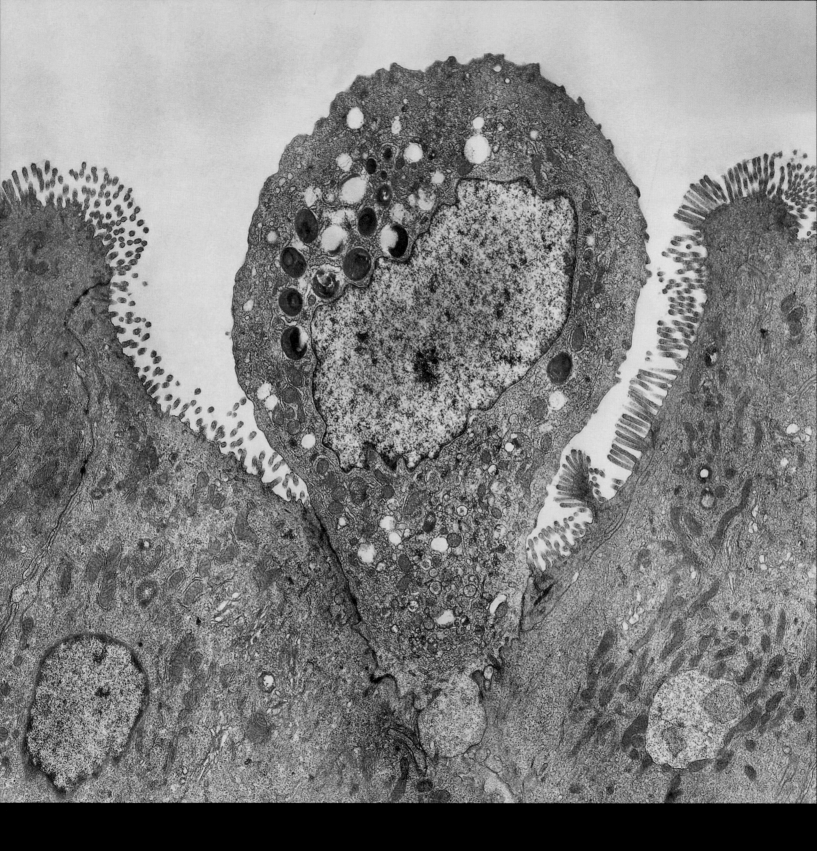

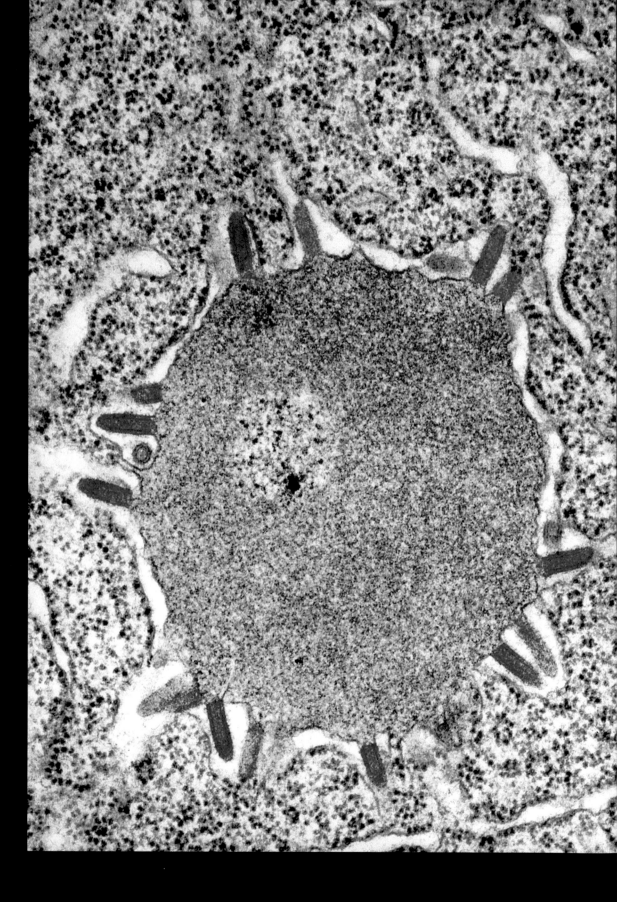

**Left: Vesicular stomatitis virus
(transmission electron micrograph)**

Vesicular stomatitis is a disease of farm animals that can also infect humans, with a range of symptoms from fever to mouth blisters. It is spread by insect bites. There is no specific cure but the disease usually passes after a couple of weeks. The virus's bullet shape is typical of its family, the rhabdoviruses, which also includes rabies. Variations of the virus have been shown to destroy cancer cells and those cells infected with HIV and ebola virus. (Magnification unknown)

**Right: Lagos Bat virus
(transmission electron micrograph)**

Lagos bat virus (LBV) is another virus from the family that also gives us rabies. Here, the virions (pink) exhibit the typical bullet shape of the virus. They are attached to an intracytoplasmic inclusion body (yellow), a mass of protein generated by the virus within the infected tissue (blue). All cases of LBV have been confined to Africa, except in one French fruit bat, which had been imported from that continent. The virus has never been documented in humans. (Magnification unknown)

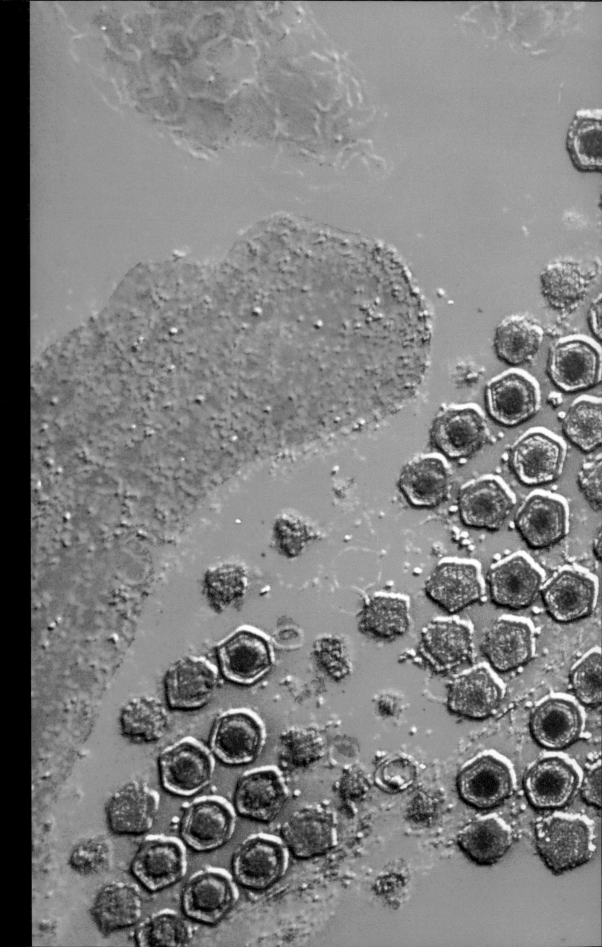

**Virions causing African swine fever
(transmission electron micrograph)**
These pink and red geometric forms are virus
particles that cause African swine fever. The
disease is present without symptoms in many
wild swine species, but is lethal in domestic
pigs. It is unusual among viruses in reproducing
not within the nucleus of the host cell but
in so-called virus factories in the cytoplasm,
which surrounds the nucleus. In the second
half of the twentieth century it spread from
Africa to Europe via the Iberian peninsula.
(Magnification x5,800 at 3.5cm)

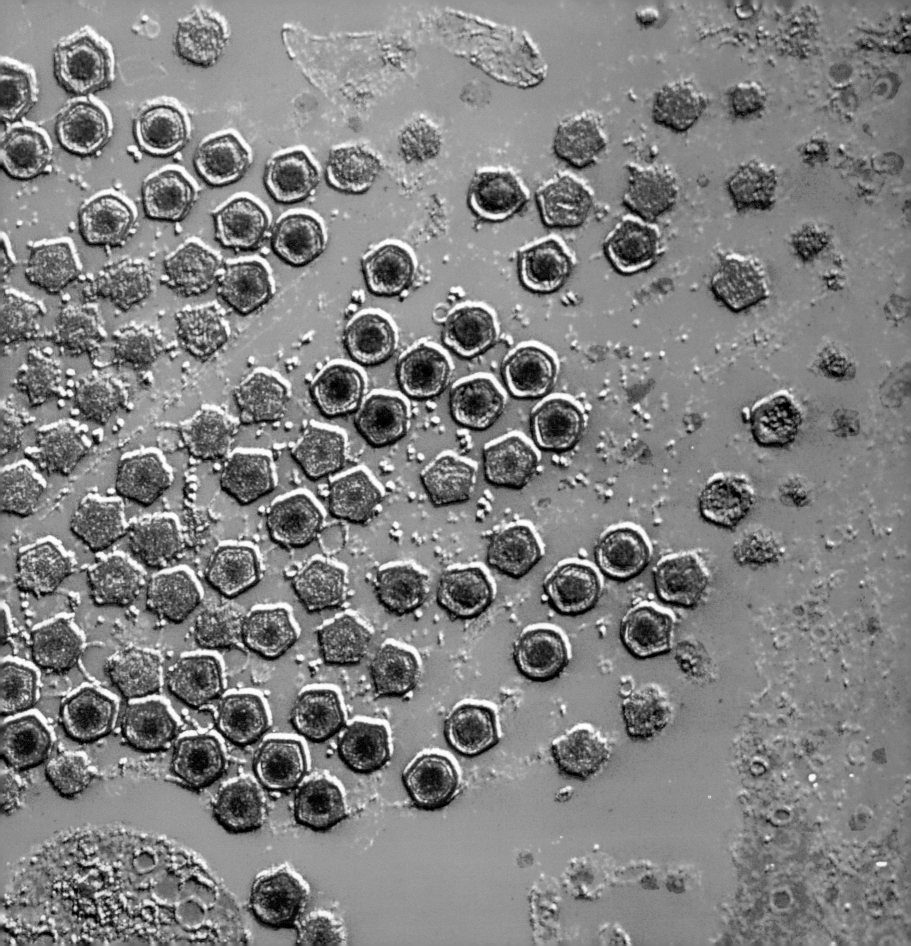

Below: Vesicular stomatitis virus (transmission electron micrograph)
Vesicular stomatitis is a disease of cloven-hoofed domesticated animals, with symptoms similar to foot-and-mouth disease – lesions of the mouth, udders and feet. As with foot-and-mouth, farmers depends on biosecurity measures to stop its spread to neighbouring farms. Although most animals recover from both diseases, mass slaughter is sometimes considered the most expedient measure of controlling it. In this image, bullet-shaped virus particles (green) are preparing to leave an infected cell (blue) after having reproduced within it.
(Magnification unknown)

Right: Bladder stone (scanning electron micrograph)
This is a crystal 'stone' recovered from the bladder of a dog. Such objects are formed by the crystallization of excess calcium oxalate, a mineral in urine. They can grow to several centimetres in diameter, causing considerable discomfort, particularly if they pass into the urinary tract. Larger ones may have to be broken up with ultrasound, or surgically removed. This one is only about 8mm in size.
(Magnification x10 at 10cm wide)

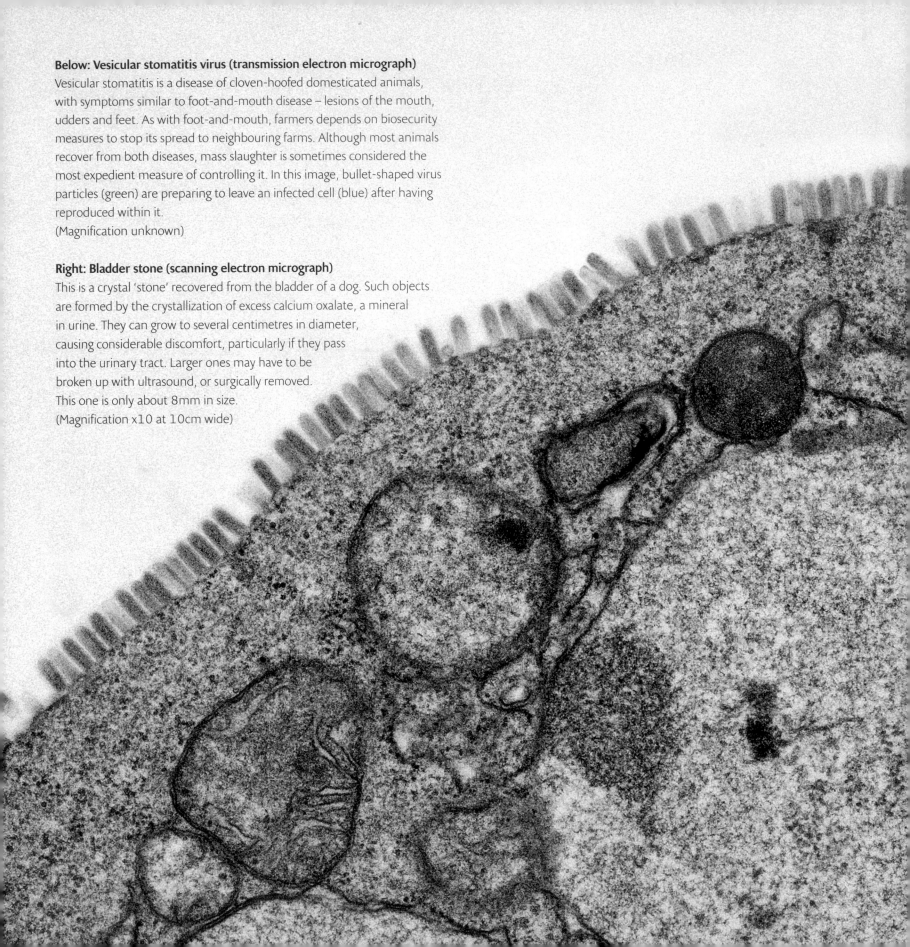

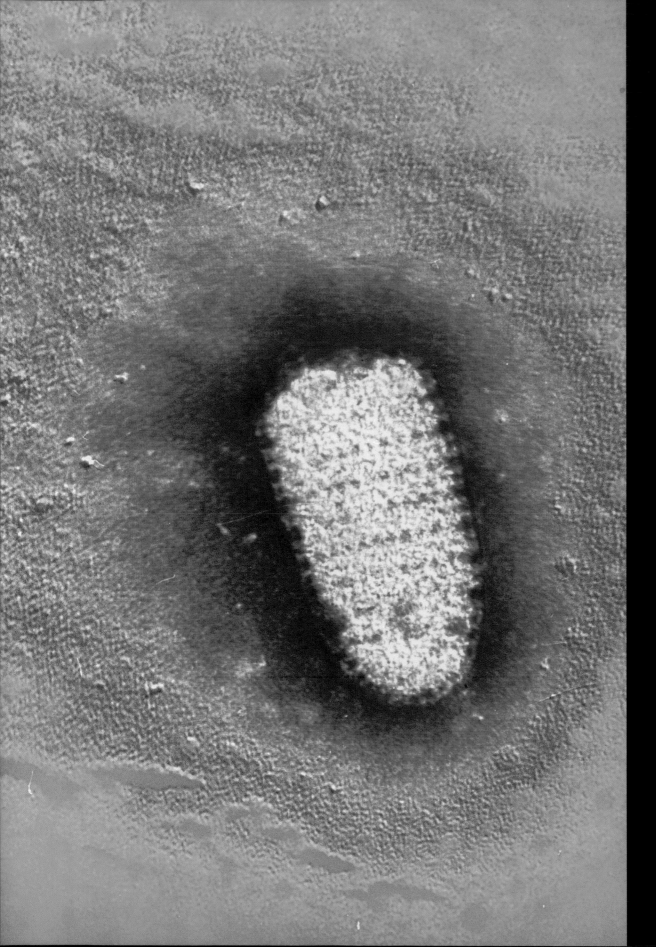

**Rabies virus
(transmission electron micrograph)**
Rabies virus is one of several lyssaviruses, all of which have this characteristic bullet shape. The virus's genetic material, ribonucleic acid (RNA), is wrapped in a protective shell called a capsid (yellow in this image). Surrounding that (in red) is the viral envelope, a mass of protein that allows the virus to bind to the host cell before infecting it. Lyssa, from which the genus gets its name, was the Greek goddess of frenzied madness.
(Magnification x100,000 at 10cm wide)

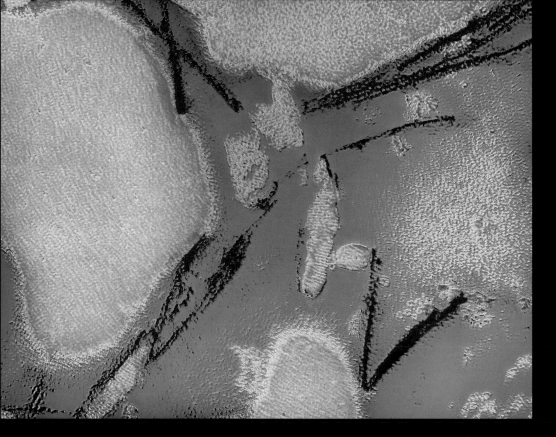

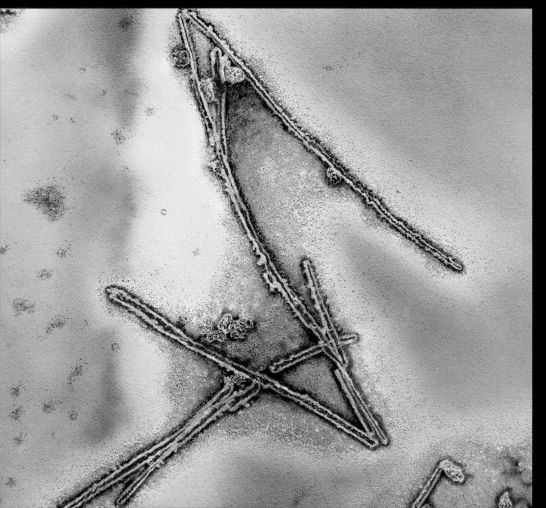

**Scrapie fibres
(transmission electron micrograph)**
Scrapie is an incurable brain disease affecting
sheep and goats. It gets its name from the
compulsion of infected animals to scrape their
fleeces off to ease the itch or inflammation
brought by the disease. It is transmitted not
by bacteria or a virus but by prions – self-
replicating proteins with no nuclei. These
red scrapie fibrils (fibre-like assemblies)
are believed to be clusters of the proteins.
Similar prion-spread diseases include bovine
spongiform encephalopathy (BSE) and
Creutzfeldt-Jakob disease.
(Magnification x58,000 at 6cm wide)

Helicobacter bilis bacteria (scanning electron micrograph)
These pasta-like spirals are a group of *helicobacter bilis* bacteria. The thin white tails growing from them are flagellae, whip-like extensions of whichfunction is mainly to move the bacteria to their destination (like the tails of sperm). The flagella may also act as sensors, seeking out the right cells for the bacteria to attack. *H. bilis* infects the intestines and livers of cats, dogs and rodents, causing hepatitis. It has also been found in humans.
(Magnification x16,500 at 10cm wide)

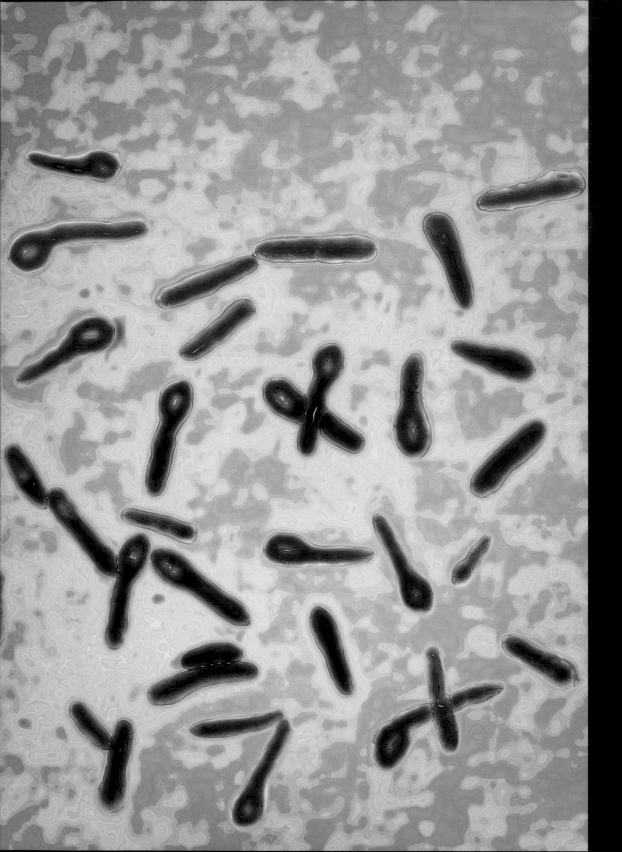

Left: *Clostridium tetani* bacteria (light micrograph)

Infection of an open wound by *Clostridium tetani* bacteria, the tiny blue tennis rackets in this image, gives us tetanus. The condition was historically known as lockjaw, and is characterized by frequent spasms that start in the muscles of the jaw but eventually progress to the whole body. The spasms may be violent enough to break the patient's own bones. The cause is a toxin produced by the bacteria which enters the central nervous system. (Magnification unknown)

Right: *Pasteurella* bacteria (scanning electron micrograph)

Pasteurellosis is a human infection caused by the bite of a pet, which introduces *Pasteurella* bacteria into the wound it makes. There are over twenty different species of *Pasteurella*, the commonest being *P. multocida*, which inflames the area around the wound and can give joint pain. It may also progress to the respiratory tract, and can cross the blood-brain barrier, causing meningitis. Penicillin antibiotics are effective at an early stage. (Magnification x17,000 at 10cm wide)

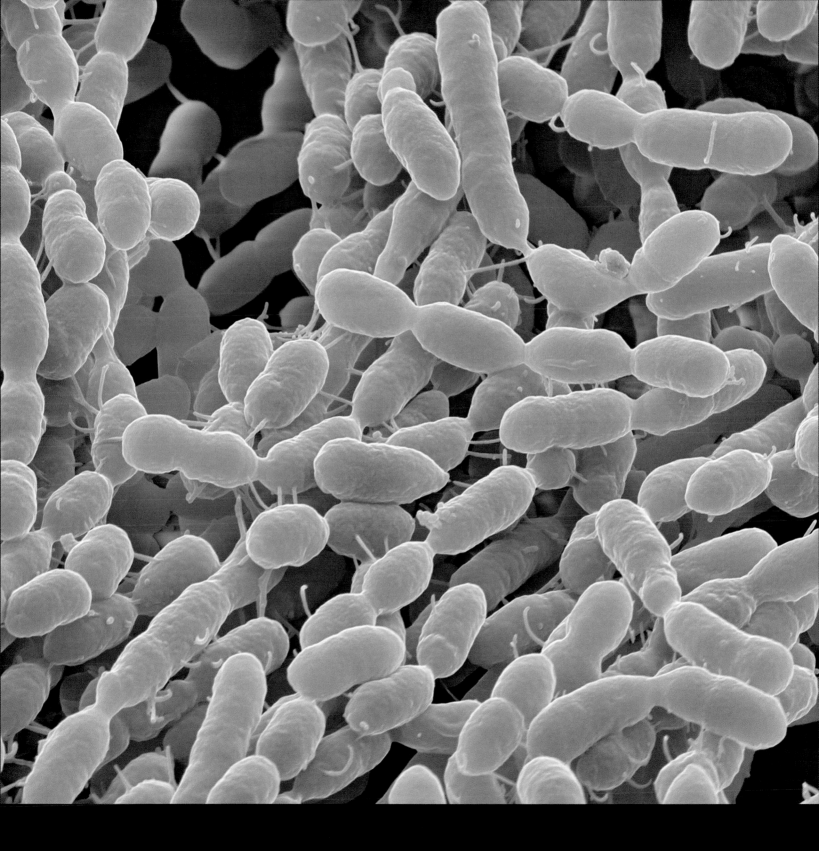

Index

Picture Credits